GUIDE TO WORLD RADIO

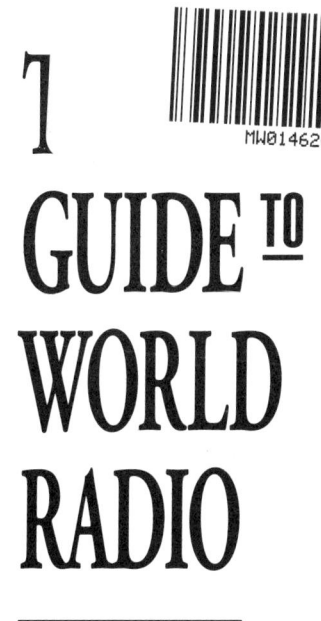

1993 EDITION

TRAVELER'S GUIDE TO WORLD RADIO

1993 EDITION

BILLBOARD BOOKS
AN IMPRINT OF WATSON-GUPTILL
PUBLICATIONS/NEW YORK

Editor: Andy Sennitt
Associate Editor: Bart Kuperus
Contributing Editor: Jonathan Marks
Art Director: Areta Buk
PostScript Programming: Marshall Ewig
Publisher: Glenn Heffernan

Copyright © 1992 by Billboard Books

First published 1992 by Billboard Books,
an imprint of Watson-Guptill Publications,
a division of BPI Communications, Inc.,
1515 Broadway, New York, NY 10036.

All rights reserved. No part of this publication may be
reproduced, stored in a retrieval system, or transmitted,
in any form or by any means—electronic, mechanical,
photocopying, recording, or otherwise—without written
permission of the publisher.

Manufactured in the United States of America

TABLE OF CONTENTS

HOW TO USE THE *TRAVELER'S GUIDE*	vii
SURVEY OF PORTABLE RADIOS BY JONATHAN MARKS	xi
INTERNATIONAL PLUG AND SOCKET STANDARDS	xxiii
WORLD RADIO BROADCASTS	1

HOW TO USE THE *TRAVELER'S GUIDE*

Doing business or vacationing in a foreign land can sometimes be stressful. The English-speaking traveler misses the conveniences of home, especially the simple things such as being able to listen to the news or sports on the radio. Many hotels around the world now provide English language satellite TV services such as Cable News Network, but for the practical traveler a portable radio is still essential.

Often, in cities where English is not the main language, there may still be local radio programs in English, or a station serving English-speaking military personnel—if only you know where and when to tune in. On the other hand, in English-speaking cities the large number of stations on the dial can prove confusing for the visitor.

If your radio can receive shortwave broadcasts, you have the additional options of various English language international services such as the Voice of America and the BBC. In some cities, these services are relayed in whole or in part by local AM or FM transmitters, or by direct satellite transmission to local cable systems. But in most parts of the world shortwave is still the only option, and to tune in these broadcasts, you need to know the times and frequencies.

That's why we produced this book. Inside, you will find the essential radio information you need in easy-to-use graphic format. At any one of the 55 major travel destinations listed, you can see exactly what English-language broadcasts are available on AM, FM, satellite and international shortwave.

GRAPHIC FORMAT

The hours of English-language broadcasts are indicated by a horizontal bar for each frequency used. The appearance of the bar varies according to whether the broadcast is daily, or only on certain days of the week, as follows:

- ▬▬▬ = Daily
- ▭▭▭ = Monday-Friday
- ▒▒▒ = Saturday-Sunday
- ═══ = Saturday only
- ≡≡≡ = Sunday only

vii

TIME INDICATION

For local stations and international broadcasters, the programming times are expressed two ways. At the top of each grid is the local time—the clock time at your destination. At the bottom of each grid is UTC (Universal Time). The UTC times are the ones that you will normally hear mentioned on the air when you listen to international broadcasters. The BBC announces times in Greenwich Mean Time (GMT), but it means the same thing.

All time indications use the 24-hour system, to be compatible with international airline and rail timetables. For example, 1500 means 3 p.m., 1800 is 6 p.m., and so on.

For local stations, the local times remain valid regardless of whether standard or daylight saving time is in use. Many international broadcasters disregard daylight saving time, however, which means that when DST is in force at your destination, you will have to tune in to these stations one hour later that the indicated local time.

PROGRAMS

For local stations, programming notes and times of news bulletins (wherever this information is available) are given underneath the station name. For the U.S. cities, we have only included those stations broadcasting news, sports and fine arts programming.

For the major international broadcasters, details of news, current affairs and business programs are given below.

BBC WORLD SERVICE

There is a BBC news broadcast every hour on the hour, varying from a one minute summary to a full hour of in depth reports and analysis. The BBC's major news programs are as follows:

Newshour: 60 minute programs at 0500, 1300 and 2100 UTC.
Newsdesk: 30 minute programs at 0000, 0200, 0700, 1100 and 1800 UTC.
The World Today: 15 minutes devoted to a single topic at 1645 and 2009 UTC (Mon-Fri) repeated at 0615 (Tues-Sat).
Worldbrief: Weekly 15 minute roundup at 2315 UTC (Fri) repeated at 0445 and 0915 UTC (Sat).

Business programs:
World Business Report: 0909, 1705 and 2305 UTC (Mon-Fri), 0905 (Sat).

World Business Review: Weekly summary and preview at 0905, 1705 and 2305 (Sun).

Additional business reports are carried at approximately 25 minutes past the hour in Newshour and most editions of Newsdesk.

Other programs:

Letter from America: The famous long-running weekly commentary on American life and current events by Alistair Cooke is on the air at 1015 UTC (Sat) repeated at 0615, 1645 and 2230 UTC (Sunday).

VOICE OF AMERICA

The VOA's English output is more regionalized than that of the BBC, and also includes some material in "Special English"—standard English spoken more slowly than normal and with a restricted vocabulary. However, there is a full ten minute news bulletin every hour on the hour in all transmissions. VOA's programs are not intended for US citizens, and the lack of comprehensive sports coverage causes particular frustration to the American abroad.

CHRISTIAN SCIENCE MONITOR

On Mondays to Fridays, the World Service of the Christian Science Monitor presents a slickly produced package of news and analysis which is likely to appeal more to the American traveler than the Voice of America's output. At weekends, apart from news bulletins on the hour much of the airtime is devoted to programming with a strong religious element.

RADIO AUSTRALIA

The output of Radio Australia can best be described as rather informal. Its world news coverage is good, with bulletins on the hour, its Asian and Pacific coverage even better. But several rounds of budget cuts have forced the station to fill an increasing amount of airtime with music, and to restrict its official target areas to Asia and the Pacific.

RADIO NETHERLANDS

On the half hour there is a world news bulletin followed (except on Sundays) by a 15 minute current affairs program called Newsline. Special versions of Newsline are beamed to Africa and Asia respectively, concentrating on events in those continents.

Radio Netherlands tries to reflect the whole spectrum of Dutch opinion, and enjoys a well deserved reputation for high quality programming.

TRAVELING WITH A PORTABLE RADIO

We have already mentioned that a portable radio is essential for the traveler. Airports obviously think so too, because their stores sell a great many of them. Nevertheless, there is a constant need for high levels of security at airports and on board flights. Unfortunately, there are no internationally recognized procedures for carrying electrical items, and rules vary from airport to airport, airline to airline, and even from route to route with the same carrier.

The only sure way to avoid hassle and inconvenience is to telephone your airline before you pack your bags, and ask what procedures apply on your particular route. Some airports and airlines request that radios be carried in hand baggage and produced for inspection, others prefer them to be in your checked baggage.

We urge all readers of this book to cooperate with the check-in and security staff, who are only following instructions to ensure the safety of yourself and your fellow passengers. Above all, do not attempt to make jokes concerning security. At least one airport now imposes heavy fines on persons found guilty of such behavior.

SURVEY OF PORTABLE RADIOS
compiled by Jonathan Marks

The continued conflicts in the world have once again highlighted the importance of keeping in touch with news events. And to do that properly requires a good quality radio. It is easy to pick up a US$20 radio, but the performance is so disappointing it is simply money wasted. The last 12 months has seen several new portable radios launched onto the world market. The "Travelers Guide" team have purchased dozens of samples at shops in Europe, the Middle East and North America. For this survey we've limited ourselves to sets costing under US$500. Sets above this price range are no longer 'portable', and offer extra features that the average traveler would not need. The fact that the selection falls amongst five manufacturers simply demonstrates that in the portable class, the choice of good sets is limited.

WHERE TO BUY

Buying a good radio for international radio listening is easy. Availability is no longer the problem, although sales staff are rarely able to offer much in the way of objective advice. Weekend editions of newspapers such as the "New York Times" often carry adverts from photo-discount outlets. Prices quoted their for receivers may be as much as 10% lower. Before purchasing by mail-order, as opposed to a local dealer, check that you are comparing like with like. Is the set being sold with a AC power supply? What are the warranty terms? Does it have an English language instruction manual? Is the set really in stock? A little research may save a lot of money.

Take the question of price difference for instance. Sales staff often give all sorts of reasons for why one set is US$150 more than another. In fact, it has nothing to do with the radio being "more powerful" And in many cases an expensive radio is not more "sensitive". The telescopic antenna on every portable radio picks up all kinds of radio energy and feeds it into sensitive circuitry inside the set. The radio therefore has to cope with all these incoming signals and select the chosen station. Cheap radios tend to get somewhat mixed up, with the result that strong signals break through onto parts of the dial where the stations concerned are not really broadcasting.

EASE OF TUNING

Tuning convenience is also important. Back in 1980, the Sony Corporation of Japan launched one of the first travel portable shortwave receivers with keypad tuning designed specifically for international broadcast reception. Until that point, conventional designs had offered one or two shortwave ranges. But entire shortwave broadcast meter bands were compressed into a fraction of an inch on the dial. If you so much as breathed heavily, the radio danced onto the next station.

There are currently some 80 travel portables on the market. Prices of travel portables are cheapest in the US....you'll pay much more in Europe and even most of Asia. We have examined them all and selected the best. Our tests are completely independent of the manufacturers and based on more than 25 years of experience in this field. The radios are divided into two groups.

- *"Pointer-and-Dial" traditional designs.* They are relatively cheap, yet provide good results. They are not as easy to tune as the more expensive sets in the digital portable class. If you want to tune in 6165 kHz on shortwave for instance, you will have to put the pointer around 6.2 on the dial and carefully adjust the dial until you hit on the right station. International broadcasters on shortwave tend to make maximum use of the available space, so stations are close together. After our survey, we can only recommend two radios of this kind, one of which is now discontinued. There are a lot of radios built in Taiwan and the Peoples' Republic of China available under a variety of brand-names. But their design is so simple that strong stations simply overload the sensitive circuitry.

- *"Digital Keypad" tuning.* These radios have a calculator style keypad. If you want 9590 kHz you tap in 9 5 9 0 on the keypad and the radio locks into that channel. Naturally you pay extra for this feature, but it saves a lot of time. Such sets also have memories allowing you to store the frequencies of your favorite stations. If you need business news from the BBC for instance you can switch on seconds before the program without having to spend five minutes searching the bands again.

RADIOS TO AVOID: TESTED & REJECTED

We examined and rejected the following models from this Receiver Survey because the short-wave coverage was obviously an extra: Audio Sonic TKS-326, TK-333S, TKS-342, TK-344F,

TKS-350, TKS-354, Grundig Music Boy 170, Grundig Prima Boy 70, Grundig Boy 40, Grundig Concert Boy 230, Grundig Cosmopolit, Panasonic GX80, GX50, GX30. Philips D 2615, Philips D2225, Philips D8184, Philips D8478, Sony AIR-7, Sony AIR-8, Sony SW15, Sony SW20, Sony WA-6000, Supertech WE-9, Supertech WE-110A.

Performance between 3 & 30 MHz was substandard, or the radio did not have enough shortwave bands to be complete. Some manufacturers (e.g. Sony) have brought out new models without the new 22 meter band (13 MHz). The new Sony ICF-SW15 for instance is therefore unable to pick up some of the clearest signals from Canada, Holland, USA, Austria, Australia, etc.

We also rejected all models with the brand name "Marc", "Magton" & "Tokyo Skylark", "Yoko" & "Frontech".

POINTER-AND-DIAL SETS

GRUNDIG YACHT BOY 204/206

This receiver has been designed by Grundig, although this appears to be the first time the German company has gone to a Taiwanese company to have it manufactured. Product Manager for Audio Manfred Lichius explained to us that they want to keep a low-cost analogue portable in their line-up, despite the fact that prices for digital portables with a synthesizer can now be made for the same price. Analogue sets still offer a much lower background noise, and generally better performance.

The Yacht Boy 204 is identical to the 206, except that the more expensive 206 has a digital clock and timer built-in. You can also set the radio to switch off after a preset period, which is useful if you want to hear music to send you to sleep.

The 206 offers no less than 15 bands, a lot more coverage than many sets in the same price bracket. Medium & long wave are available, plus FM in mono. Short-wave coverage of the 120, 90, 75, 60, 49, 41, 31, 25, 22, 19, 16, and 13 meter bands is provided, including plenty of coverage either side of the "official"

bands. The set weighs 1.2 lbs, including the three AA cells needed to power the radio, and the single AA battery used for the clock/timer. A plastic stand on the back of the set allows the radio to be tilted when operated on a table. If the stand is accidentally hit (e.g. by accidentally dropping the set) it is designed to pop out without snapping off. Replacement is easy.

Power consumption is remarkably low, averaging around 25 mA at a comfortable listening volume, even on FM. Band spread on the receiver is good, the tuning knob having just a slight backlash. Despite the single-conversion design, the image rejection is quite good for a receiver of this type. Listening in the evening hours on the 49 meter band there was surprisingly few "ghost" signals.

At a price of around US$70 in Europe, this set offers excellent value for the traveler looking for a no nonsense short-wave radio. In North America the set is more expensive, retailing at around US$120. Since the Sony ICF-7601 is also retailing at around the same price a comparison of these two is worthwhile. The Sony gives better short-wave performance (thanks partly to its dual conversion design) but you may prefer the fuller audio of the Grundig. More information from Grundig, Kurgartenstrasse 37 Fuerth/Bayern D8510, Germany. Tel: +49 911 7030. In the USA, Grundig has a distributor. This is Lextronix Inc, 3520 Haven Ave, Unit L , Redwood City, CA 94063. Tel: +1 415 361 1611 or in the USA only call (800) 877 2228. In Canada call (800) 637 1648.

SONY ICF-7601

This is Sony's cheapest 'paperback-book-size' portable radio, actually released onto the world market in March 1988. It is important to check the model number is exactly ICF-7601. Dealers may try to sell you a similar sized radio called the ICF-7600A or ICF-7600DA both of which are different radios and not up to the same standard. The ICF-7601 is priced at around US$119.95 in the United States. It weighs 1.2 lbs without the batteries, so it is therefore quite light.

Mediumwave and FM (in mono) are standard. Then come the 9 shortwave broadcast bands, i.e. the 13, 16, 19, 22, 25, 31, 41, 49 and 60 meter bands.

Finally on European models designated the ICF-7601L coverage of the longwave band is offered. On models sold elsewhere, a scale covering the broadcast bands of 75, 90, and 120 meter bands is provided. The 90 and 120 meter bands are reserved for stations in the tropical regions of the world.

The set copes quite well with strong shortwave signals. A two-position tone control helps to reduce some of the familiar shortwave whistles when it is set to the news position. If you are someone who likes to listen in subdued lighting, then the lack of a dial light might bother you.

Battery consumption on the four penlight cells is slightly above average. In Europe a 6 volt DC mains power supply is included in the price and works out far cheaper to use than alkaline batteries. Rechargeable batteries would be another option if you are forced to travel light. A lock switch is provided so that the receiver will not accidentally switch on while packed in a suitcase.

Overall the Sony ICF-7601 offers excellent value.

DIGITAL PORTABLES

PANASONIC RF-B45

Launched in North America in early 1991, this receiver is clearly based on its predecessor, the RF-B40, with improved fine-tuning and the additional capability of single-sideband reception. What they've done is to take off some of the features seen on the RF-B65L and build it into a smaller case. It has an illuminated on-off button so you can see at a glance if the radio has been left on with the volume turned down.

On most versions, the set covers longwave between 146 and 288 kHz, mediumwave between 522 and 1611 kHz, and provides continuous shortwave coverage between 1611 and 29995 kilohertz. VHF FM between 87.5 and 108 MHz is also offered. On models sold in Italy the coverage of the receiver is partly restricted in the shortwave range. The RF-B45 can be tuned in one of two ways. On the front is a block of 12 dark-gray keys, similar in layout to a pocket calculator. Ten of the buttons correspond with the digits 0 to 9, there's one for the decimal point, and a button for storing chosen frequencies in the memory. If you wanted to select 9895 kHz, you press a button marked frequency, and then tap in 9 8 9 5 or 9 . 8 9 5 and push the enter button. Should you push 9 8 . 9 5 the receiver automatically assumes you mean 98.95 MHz FM, and that indicates the use of very user friendly software in the micro-processor stage.

Each of the 12 push buttons on the keypad also has a small text on it, corresponding to the 12 different broadcast bands between 2 and 30 MHz. If you're on your travels and can't recall where the 25 meter band starts for instance, you simply push a button marked meterband, and by pressing key number 8 on the panel, the set springs to 11650, i.e. the lower frequency end of the 25 meter broadcast band. Manual tuning is offered too. On SW, you press a key and the set jumps by 5 kHz to the next channel. If you keep the button pressed, the receiver continues to move up the dial at about 50 kHz a second. On medium wave you can select whether the receiver moves in either 9 or 10 kHz at a time. On longwave the steps are 9 kHz and 50 kHz on VHF FM. But note that the set mutes as it scans.

Fortunately, unlike the old RF-B40, the new RF-B45 model has a fine-tuning control. This takes the form of a thumbwheel on the side of the receiver which adequate for most broadcast listening. Resolving SSB signals is more difficult, but not impossible. The set offers 27 memories, rather than 36 offered on the RF-B65.

At a price of around US$170.00 this set offers a strong competitor to the Sony ICF-SW7600, and curiously to the Panasonic RF-B65 as well. The advantages are ease of operation, size, weight, and good audio quality.

PANASONIC RF-B65L

Another "paperback book", weighing in at 1.7 pounds including the 6 penlight batteries that fit inside. When switched off, the digital liquid crystal display shows the local time, or you can set it to show a second time, e.g. UTC. The set covers FM between 87.5 and 108 MHz, longwave between 155 and 519 kHz, medium wave between 520 and 1610 kHz, and shortwave between 1615 and 29999 kilohertz. Some versions of the RF-B65 are being sold in parts of the world without longwave.

In addition to keyboard tuning, there is a conventional rotary tuning knob. In terms of size, weight, and facilities this set is clearly designed to compete with the SONY ICF SW-7600 portable radio. This Panasonic model is virtually identical to the Sony when it comes to sensitivity, and selectivity (the ability of the set to pick out the station you want from the interference). Accurate tuning is easier though, as the set moves up and down the bands in finer steps, i.e. 1 kHz.

The set has quite a high drain on the 4 penlight batteries in the radio section. Listening 3 hours a day, the batteries won't last much longer than about week. An optional 6 volt DC power supply is available to allow you to use the set off the household current supply.

The receiver costs around US$220. Bearing in mind the similar performance to the Panasonic RF-B45, the Sony ICF-SW-7600 and Sangean ATS-808, check the prices of these competing receivers before you make a final selection.

SANGEAN ATS-808

With a price tag of around US$200, this set is designed to heavily compete with other "paperback book" size digital portables. The ATS-808 measures just 7.7 × 4.9 × 1.4 inches, and weighs 1.53 pounds including the 6 penlight batteries (two of these are used to drive the on-board computer, the other four to run the radio). It is marketed alongside the older (and much larger) ATS-803A for the duration of 1993 at least.

The ATS-808 is designed for simple operation. It is for the reception of broadcast signals only. However, the ATS-808 has provision for two bandwidth filters, Frequencies are shown on a LCD display which is easy to see in daylight. At night you can't see any of this information because there is no backlighting of the display. The radio has a timer, and a dual clock...that is useful if you want to keep track of the difference between UTC and local time.

The ATS-808 covers 150-1620 kHz, 2300 -26100 kHz, and 87.5-108 MHz. If you listen to the radio using headphones, then stereo reproduction on FM is possible. The tuning software has been well thought out. You can also ask the radio to select a particular meter band if you cannot remember a particular frequency. The ATS-808 offers two means of manual tuning control. Either a set of up-down buttons, or a manual rotary tuning knob at the side of the set.

The radio offers 45 memory channels, nine each on LW, MW, and FM, 18 on shortwave. In Europe, Siemens of Germany distribute the Sangean radio under their own brand label (i.e. RK-661).

The radio works best on its built-in telescopic whip. The radio copes fairly well with strong signals from local transmitters, overloading during evening hours being reduced to acceptable levels by reducing the length of whip or switching the attenuator to the LOCAL position.

In short, the ATS-808 is an excellent portable radio for the price.

SONY ICF-SW7600

The ICF-SW7600 has a logical design and operation. To tune to 9895 kHz for instance, you press the AM button, the numbers 9 8 9 5, and then the AM button again. This executes the tuning request and the receiver tunes to that frequency. The keys on the keypad are well spaced out, so there is no chance of accidentally pressing two keys at once. Both coarse and finer tuning on shortwave frequencies is desirable, and possible. The liquid crystal display is clear, showing the time while the radio is switched off and the frequency when switched on. The radio has a timer function, and can be set to operate a tape recorder. The set offers 10 memories which are programmed by pressing and holding the enter key, and then selecting one of the keys from 0–9.

The test results show good shortwave performance for a radio of this type. We noted that an external long wire antenna is NOT recommended. FM performance is fine, and the addition of FM STEREO via the headphone socket is a definite plus. The audio on the ICF-SW7600 is better than on the previous models. The set measures $7.3 \times 4.6 \times 1.25$ inches, and weighs 1 lb. 5 oz.

The radio has a well designed ON/OFF switch which can be locked. This is useful when the set is packed in a suitcase and you don't want it to spring to life when knocked about. The radio works off 4 penlight batteries. A full set of batteries gave us 18 hours of good FM reception at reasonable listening volume. The batteries should be kept in the radio at all times, even when an external power supply is connected. The ICF-SW7600 is a well designed shortwave portable receiver ideally suited to international radio listening. The price of US $220 makes it a strong competitor to the Sangean ATS-808 & Panasonic RF-B65L.

SONY SW1S/E

This is an extremely compact radio about the same size as most plastic cases round an audio compact cassette. The ICF-SW1S is just very slightly thicker. It will certainly fit in the breast pocket of a medium size man's shirt, and indeed the whole radio is designed around the need for compact electronics.

The ICF-SW1S is tunable continuously between 150 kilohertz and 29995 kHz on AM, and between 76 and 108 MHz on FM. But in some countries like Saudi Arabia and Malaysia there is no coverage on the radio between 285 and 531 kHz. The second "S" in the ICF-SW1S simply stands for "system" for it is sold as a complete package with an active antenna included in the price.

It works on 2 penlight cells which fit into the back of the receiver. They slide in fairly easily, although you need to be careful not to force the batteries.

In most countries you can buy the SW1 mini portable receiver on its own or in a plastic presentation case complete with several accessories. There is a smart power supply which you can plug into any wall socket regardless of the AC power. If it is between 100 and 240 volts it will work. If you are trying to listen in a concrete apartment block or hotel which completely shields out shortwave signals, then it might be an idea to try the supplied active antenna AN-101. However, our tests show that the efficiency of the AN-101 is deliberately kept low. Not much of the signal picked up by the telescopic whip is delivered to the built-in circuitry. Putting the SW1 on its own by the window with the telescopic whip extended produced similar if not better results in Europe than trying to use the radio inside the room with the cable trailing to the active antenna by the window. In North America and the Pacific where shortwave signal levels are low, we found the active antenna more useful, especially in an apartment. The radio on its own weighs just half a pound and fits into a shirt pocket. The radio plus the plastic case, active antenna and power supply is half the size of a brief case and weighs 3.5 pounds...seven times as much!

At a price of US$300 the Sony SW1S option is not cheap.! You are paying heavily for the small size packaging, and for the other accessories in the plastic presentation kit. The fact that the set has no finer tuning on shortwave than 5 kHz may present reception problems. The set is also marketed in some parts of the world as the SW1E, which is simply a package containing the receiver without the accessories. Our tests show that this works out to be much better value.

SONY ICF-SW55

The ICF-SW55 was launched in fall 1991, and is slightly larger than the existing ICF-SW7600. It measures $5 \times 7.5 \times 1.5$ inches, and weighs 1 pound 10 oz. But you get a lot more features. There is a familiar keypad for directly entering frequencies, and also a thumb wheel which allows you to tune up and down the bands either in 1000 or 100 Hz steps. But the front panel is dominated by a large liquid-crystal display screen. A compact world map which shows your time zone with respect to Greenwich Mean Time (otherwise known as UTC), and a signal strength meter. The display also shows you if you are on either wide or narrow AM bandwidth, or upper or lower sideband when using SSB. The set has a complex but versatile five function timer to program the radio to choose certain channels at a particular time.

Thus far it looks like Sony have taken most of the features from their popular (now discontinued) ICF-2010 table-top receiver and built them into a box the size of the SW-7600. But when it comes to memories, there is considerable improvement. You can store up to 125 favorite frequencies in the set. You do this in groups of up to 5 frequencies for each favorite station. Let's say you want to store Voice of America channels in it. First of all by using a shift button and the keypad you can get the display to show 'USA'. Then under the display are five function buttons. You can program button number one to remember 6040, button 2 to remember 9760 kHz, and so on. Imagine it rather like an electronic logbook, where each page contains the name of a station, and up to 5 favorite frequencies.

The set comes with some stations already programmed in it, although some of the frequency choices that Sony have made are rather peculiar. There is FM stereo on headphones, and continuous coverage on AM from 150 through to 29,999 kHz. The recommended price for the SW55 is around US$384. The SW55 has a new big brother, the SW77 which has a lot of extra features for around US$120 more. But the weight of the SW77 precludes it from the business traveler market.

PORTABLE COMMUNICATION RECEIVERS

LOWE HF-150
If you take international radio listening very seriously you might want to consider taking a communications receiver with you on

your travels. Until now that has meant finding a lot of space in the suitcase. But we have examined one receiver, the Lowe HF-150, which is different. Lowe is an English company which started making radios to its own design and specification a couple of years back. In October 1992 we tested an off-the-shelf example of the HF-150, putting it through a series of laboratory and practical listening tests. For a price of £329 in Britain (including VAT), you get a communications receiver in a compact box that is easy to put into a suitcase. The case is made of metal, not plastic. It's quite light too, just under 3 lbs without the 8 penlight batteries which fit into two special holders at the back of the set.

From the front there are just 5 controls....a combined on-off switch and volume control, three buttons which have several functions including the selection of the mode and memories, and a large tuning knob. A large 5 digit liquid crystal display shows the frequency you're tuned to within the nearest kilohertz, If you push a button the display gives you information about the receiver mode and memory number, but normally it shows only the frequency, and there's no light to illuminate it. That's it. Lowe sell a keypad as an optional extra that plugs into the back of the set and you place in front of the radio as you use it. That's essential if you want to move quickly about the dial...otherwise you have to move up and down in frequency by spinning the tuning knob. Getting from 30 kHz right up to 30 MHz, which represents the full coverage of the set, could take some time.

The set has no signal strength meter, you can't add extra filters at a later stage for very narrow bandwidth reception of Morse code, there's no notch filter, no noise blanker, and no tone control. But if these are extras that you can miss, then what Lowe have put inside the box turns out to be very acceptable indeed.

The dual-conversion super heterodyne design is quite straight forward. Signals come in best from an external antenna. But as this is usually difficult to mount in a hotel, you can switch in an antenna amplifier if you're using an indoor whip. Signals go through a 30 MHz low pass filter before they hit the mixing stage of the radio. This gives you a dynamic range of 86 dB which is a fair-to-good value for a radio of this price.

xxi

There's not much difference in sensitivity between the wide and narrow filters used in the HF-150. The radio has various modes. USB, LSB, standard AM, and you can also use what's termed synchronous AM. Unlike other Lowe sets available until now, the HF-150 allows you to listen to either the upper or lower sideband of a broadcast signal whilst in the "sync" mode. That's extremely useful when there's a strong interfering station 5 kHz away from the station you're trying to listen to. You can also use synchronous detection to reduce at least some of the effects of short-wave fading.

Battery consumption of the receiver is quite high, especially when compared to similar priced competition, anything up to 275 mA at full volume. We put in a set of 8 fresh alkaline batteries and got the radio to work for just 6 hours before they were flat. You can purchase rechargeable nickel cadmium batteries. When the set is switched off they automatically charge up. It takes about 16 hours to get a full charge after which you can use the radio for portable work for about 3 hours before you need to recharge again. The cheapest solution of all is simply to use the supplied external AC adapter which plugs into the back of the radio and gives all the power you need without any hum problems.

The HF-150 has two filters which have a bandwidth of 7 and 2.9 kHz respectively. These ceramic filters have a good shape factor for the price paid. So if the signal you want to listen to is strong you can really sit back and listen to the programming. The design of the automatic gain control is excellent, so no unwanted pumping of signals. The signal distortion is very low for a radio of this time, and if you connect the radio to a hi-fi set you'd be surprised what fidelity you can get out of a strong short-wave broadcaster.

Further information from: Lowe Electronics, Chesterfield Road, Matlock, Derbyshire DE4 5LE, England. Tel +44 629 580800. Fax: +44 629 580020.

WORLDWIDE PLUG AND SOCKET STANDARDS

In some countries, radio batteries are very expensive or hard to find. Airport shops and electrical stores usually sell voltage adapters which permit you to run your set from the local mains supply. A variety of different plugs and sockets are used worldwide, so we have included diagrams of each type with a list showing where they are used. Look up your travel destination on the list and compare its plug type with the diagram. Make sure you have the correct type for your destination.

A word of caution—to avoid damaging your radio or the power supply, check the voltage information in this book and make absolutely sure that the power supply is set correctly before plugging it in.

Plug diagrams and a list of cities are on the next two pages.

SOCKET PATTERN	PLUG PATTERN
A	
B	
C	
D	
E	
F	
G	

CITY NAME	PLUG PATTERN
ABU DHABI	D, G
AMSTERDAM	C, F
ANKARA	C, F
ATHENS	C, F
AUCKLAND	I
BANGKOK	A, C
BARCELONA	C, F
BEIJING	C, I
BERLIN	C, F
BOMBAY	C, D
BRUSSELS	C, E
BUENOS AIRES	C, I
CAIRO	C
CALCUTTA	C, D
CARACAS	A, B
CHICAGO	A, B
COPENHAGEN	C, K
DELHI	C, D
DUBAI	D, G
DUBLIN	F, G
FRANKFURT	C, F
GENEVA	C, L
GLASGOW	D, G
HAMBURG	C, F
HELSINKI	C, F
HONG KONG	D, G
KUALA LUMPUR	G

SOCKET PATTERN	PLUG PATTERN
H	
I	
J	
K	
L	
M	

Courtesy of Panel Components Corporation

CITY NAME	PLUG PATTERN
LAGOS	D, G
LONDON	D, G
LOS ANGELES	A, B
MADRID	C, F
MANILA	A, B, C
MELBOURNE	I
MEXICO CITY	A
MILAN	C, M
MONTREAL	A, B
MOSCOW	C, F
MUNICH	C, F
NEW YORK	A, B
OSLO	C, F
PARIS	C, E
PERTH	I
RIO DE JANEIRO	A, B, C
RIYADH	A, B, E, F
ROME	C, M
SEOUL	A, F
SINGAPORE	C, D, G
STOCKHOLM	C, F
SYDNEY	I
TAIPEI	C, I
TEL AVIV	J
TOKYO	A, B
TORONTO	A, B
VIENNA	F
ZURICH	C, L

WORLD RADIO BROADCASTS

2 ABU DHABI

ABU DHABI, UNITED ARAB EMIRATES
Hours difference from UTC (GMT): +4
Main business languages: Arabic, English
Electricity: 50Hz, 220V
Currency: UAE Dirham
Telephone country prefix and area code: 971 (2)

LOCAL AM & FM STATIONS

	FREQUENCY
Radio of the U.A.E. Government station.	8:0
Capital Radio Government commercial station.	FM 93.5

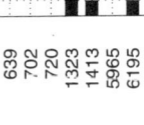

INTERNATIONAL BROADCASTERS

BBC World Service
639
702
720
1323
1413
5965
6195
9580
11760
11955

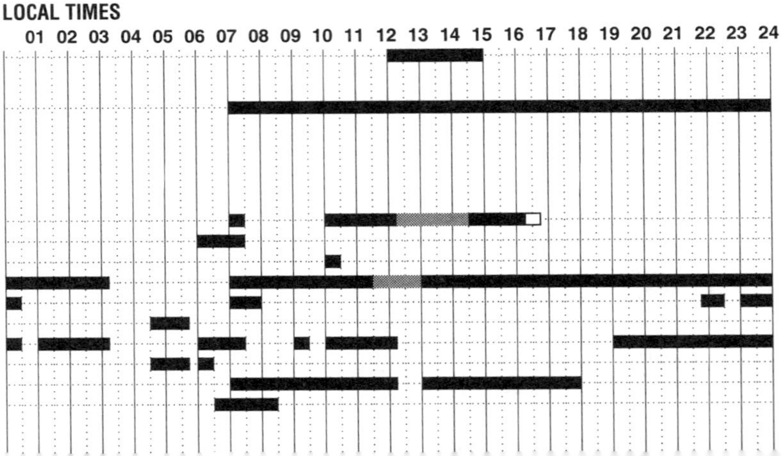

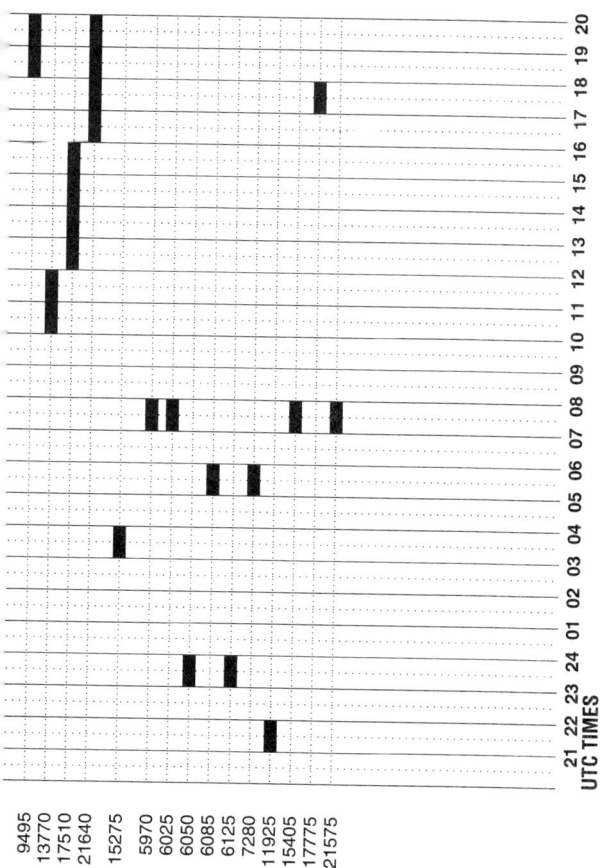

ABU DHABI 3

Station	Frequencies (kHz)
Christian Science Monitor World Service	9495, 13770, 17510, 21640
Radio Canada International	15275
Radio Japan	5970, 6025, 6050, 6085, 6125, 7280, 11925, 15405, 17775, 21575

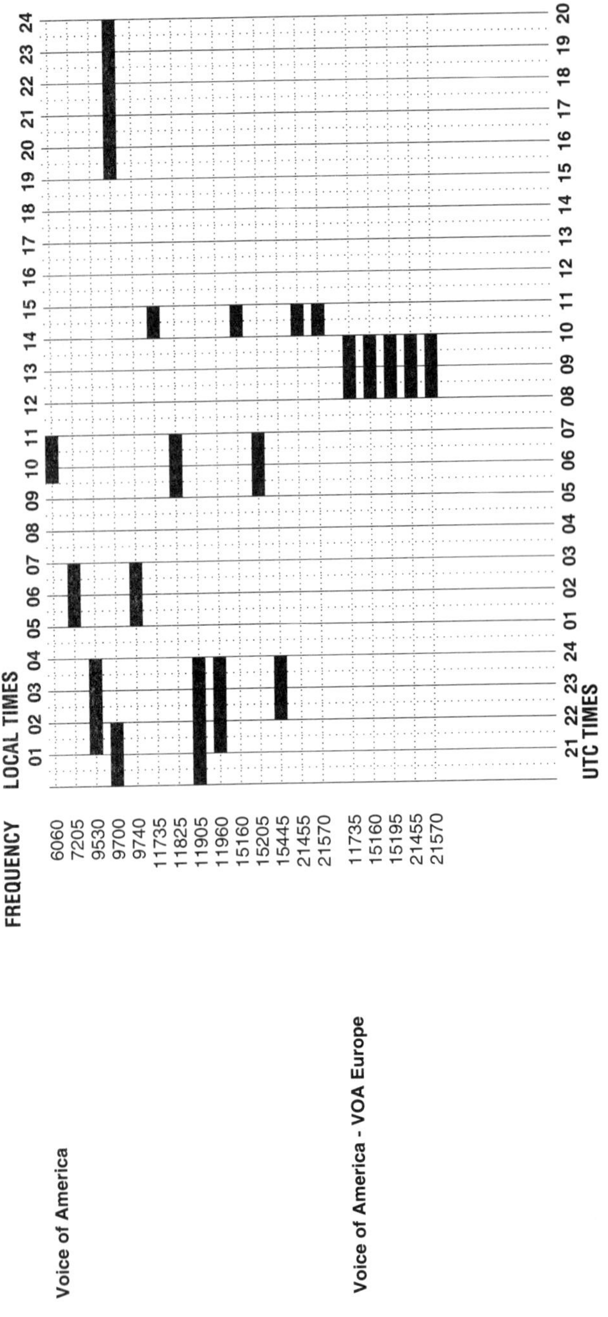

4 ABU DHABI

Voice of America

FREQUENCY
6060
7205
9530
9700
9740
11735
11825
11905
11960
15160
15205
15445
21455
21570

Voice of America - VOA Europe

FREQUENCY
11735
15160
15195
21455
21570

AMSTERDAM 5

AMSTERDAM, THE NETHERLANDS

Hours difference from UTC (GMT): +1 (Summer +2)
Main business languages: Dutch, English
Electricity: 50Hz, 220V
Currency: Guilder
Telephone country prefix and area code: 31 (20)

LOCAL AM & FM STATIONS

	FREQUENCY
BBC Radio 4 British domestic network. Main news 0800, 0900, 1400, 1800, 1900, 2300, 0100.	198
BBC Radio 5 British sports and educational network.	693 909
Voice of America - VOA Europe Program for young Europeans. News on the hour. Available on hotel cable system only.	FM 99.1

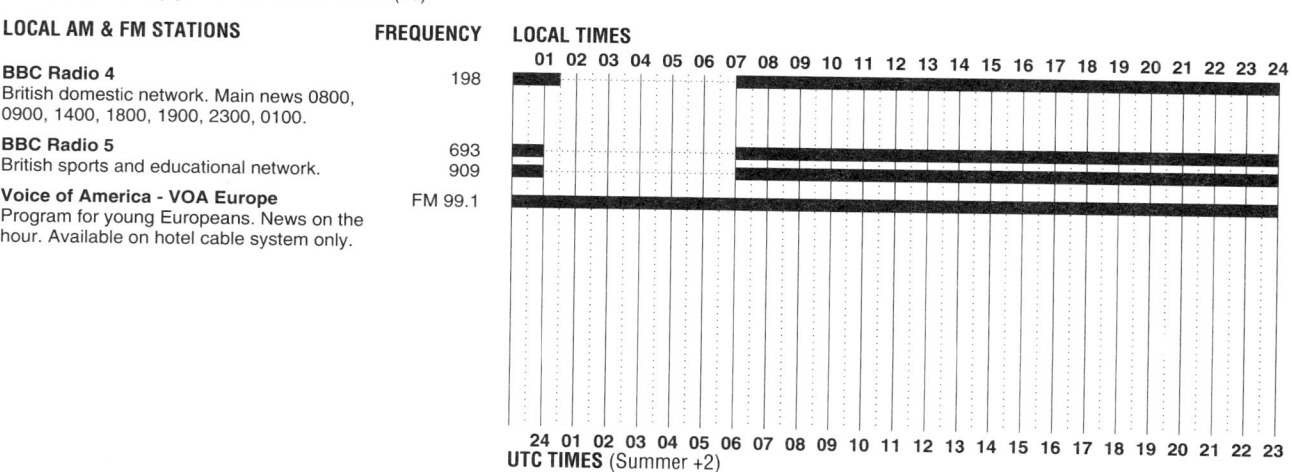

6 AMSTERDAM

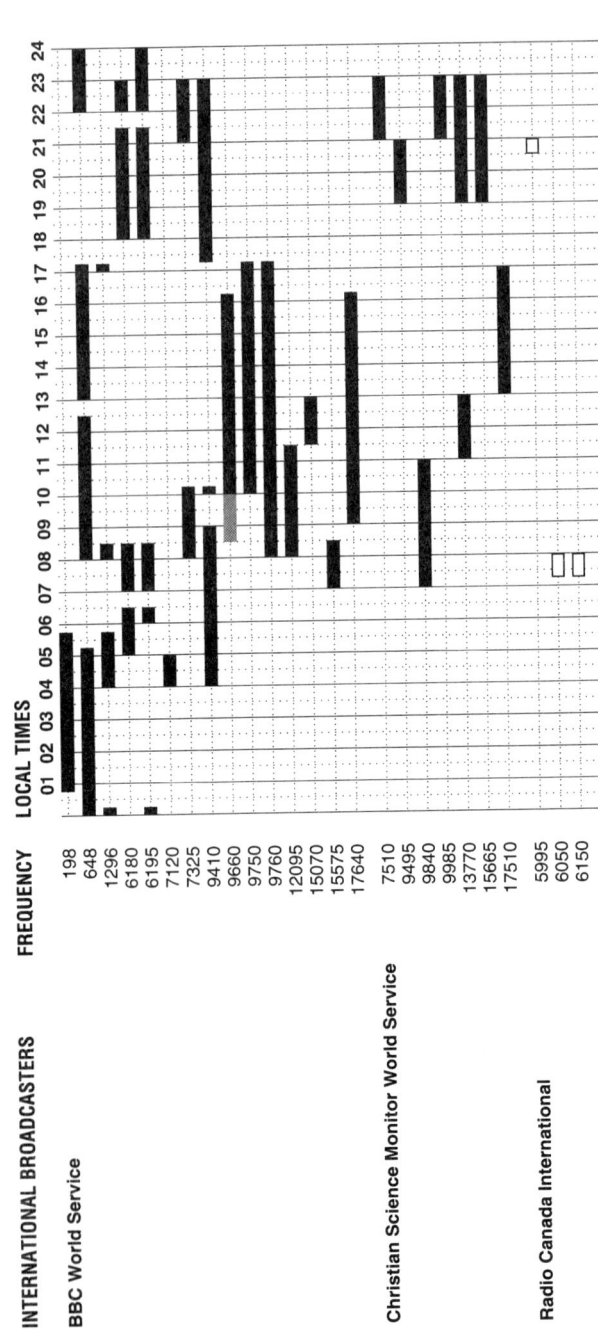

INTERNATIONAL BROADCASTERS	FREQUENCY
BBC World Service	198
	648
	1296
	6180
	6195
	7120
	7325
	9410
	9660
	9750
	9760
	12095
	15070
	15575
	17640
Christian Science Monitor World Service	7510
	9495
	9840
	9985
	13770
	15665
	17510
Radio Canada International	5995
	6050
	6150

AMSTERDAM 7

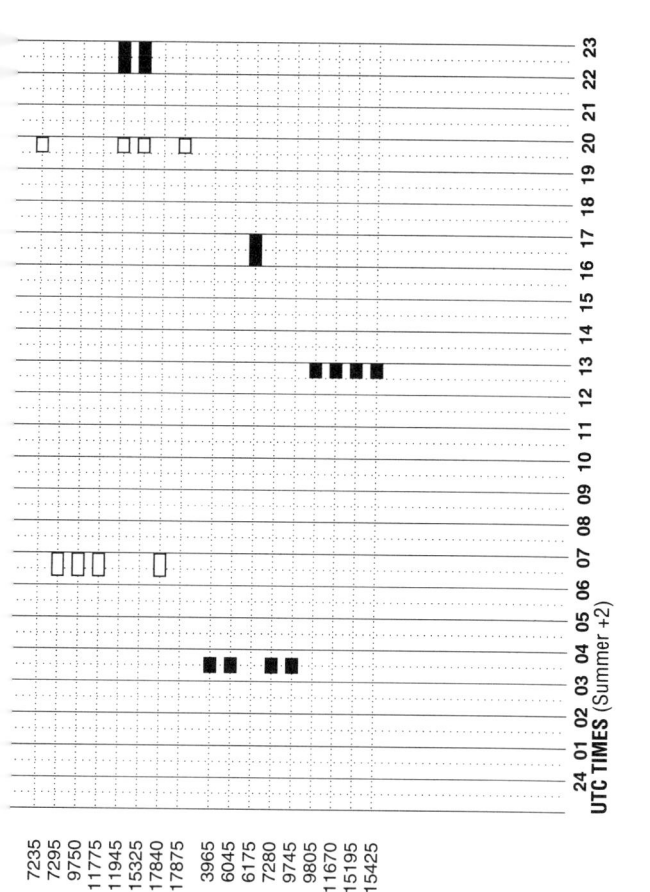

Radio France International

Frequencies (kHz):
7235, 7295, 9750, 11775, 11945, 15325, 17840, 17875
3965, 6045, 6175, 7280, 9745, 9805, 11670, 15195, 15425

UTC TIMES (Summer +2)

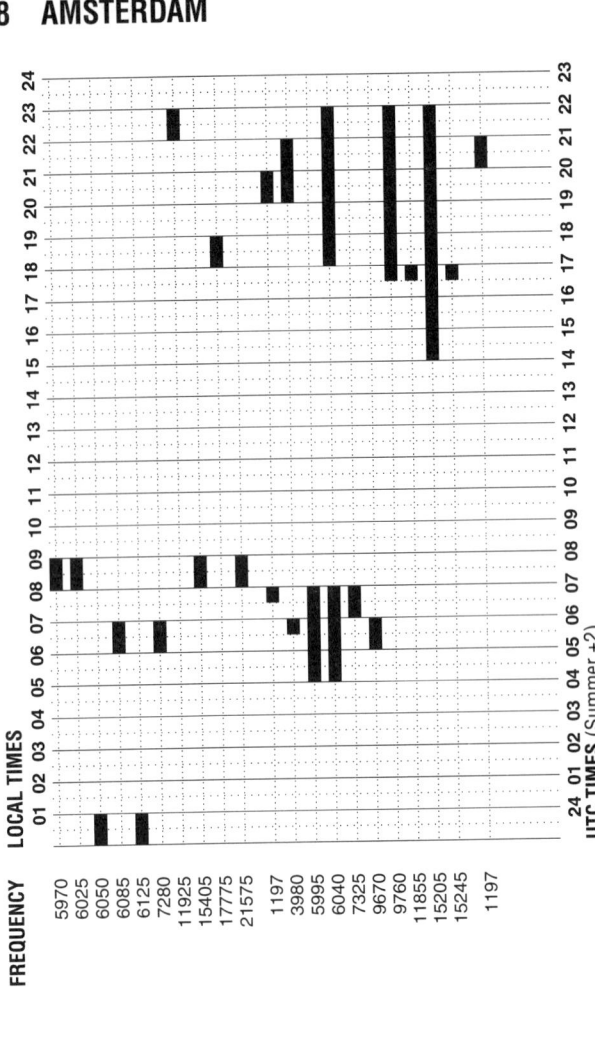

8 AMSTERDAM

FREQUENCY		
5970		Radio Japan
6025		
6050		
6085		
6125		
7280		
11925		
15405		
17775		
21575		
1197		Voice of America
3980		
5995		
6040		
7325		
9670		
9760		
11855		
15205		
15245		
1197		Voice of America - VOA Europe

LOCAL TIMES
UTC TIMES (Summer +2)

ANKARA, TURKEY

Hours difference from UTC (GMT): +2 (Summer +3)
Main business languages: Turkish, English
Electricity: 50Hz, 220/380V
Currency: Turkish Lira
Telephone country prefix and area code: 90 (41)

LOCAL AM & FM STATIONS   FREQUENCY

TRT-3   FM 91.2
Government station. News summaries 0903, 1203, 1403, 1703, 1903, 2203.

INTERNATIONAL BROADCASTERS

BBC World Service
6195
7120
7325
9410
9750
9760
12095
15070
17640

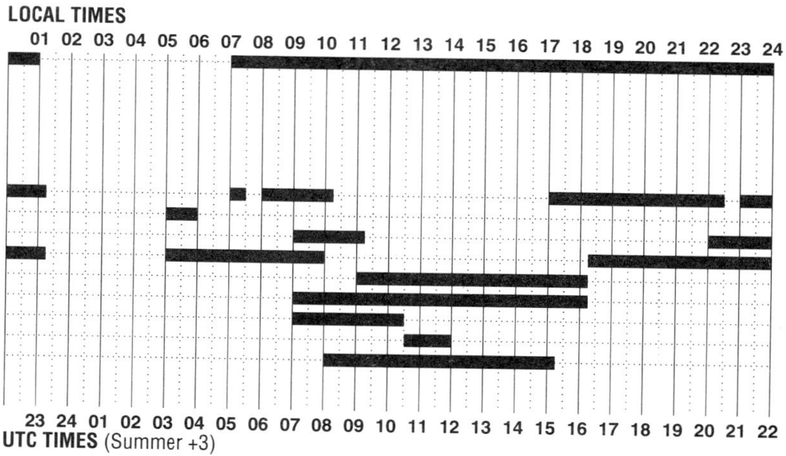

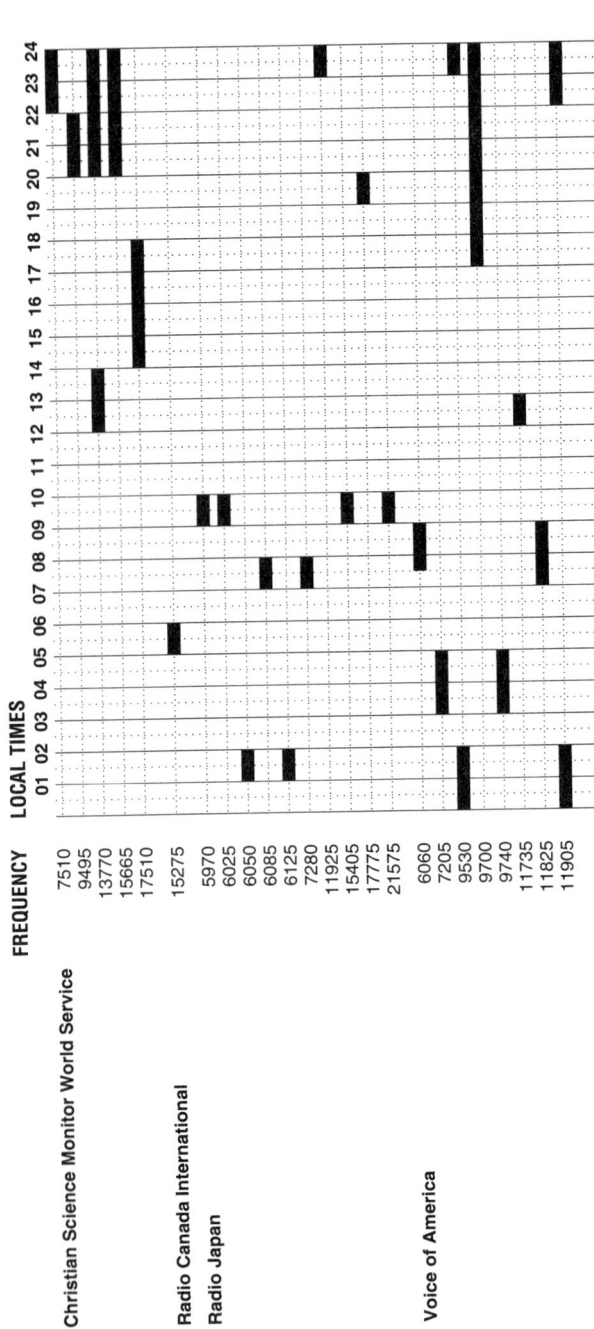

10 ANKARA

Broadcaster	Frequency
Christian Science Monitor World Service	7510
	9495
	13770
	15665
	17510
Radio Canada International	15275
Radio Japan	5970
	6025
	6050
	6085
	6125
	7280
	11925
	15405
	17775
	21575
Voice of America	6060
	7205
	9530
	9700
	9740
	11735
	11825
	11905

ANKARA 11

Voice of America - VOA Europe

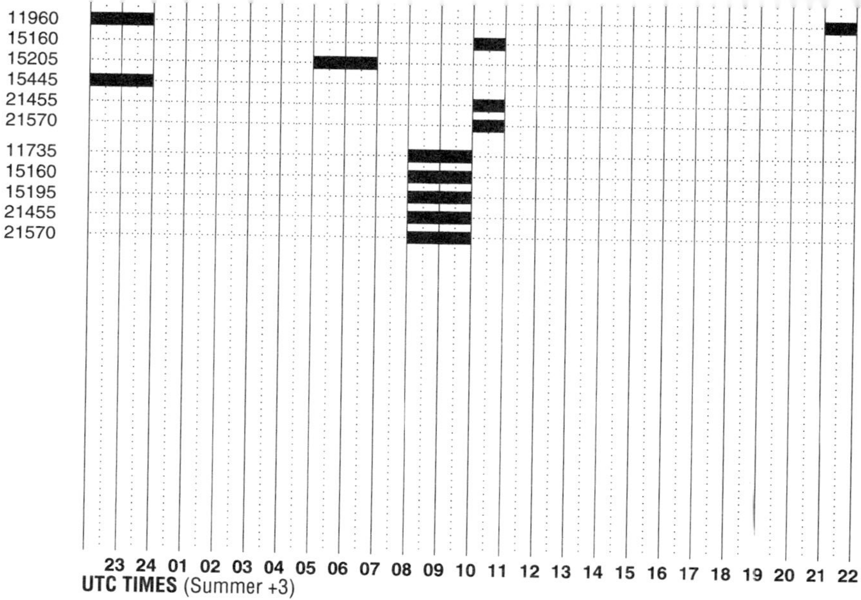

Frequencies
11960
15160
15205
15445
21455
21570
11735
15160
15195
21455
21570

UTC TIMES (Summer +3)

12 ATHENS

ATHENS, GREECE
Hours difference from UTC (GMT): +2 (Summer +3)
Main business languages: Greek, English
Electricity: 50Hz, 220V
Currency: Drachma
Telephone country prefix and area code: 30 (1)

INTERNATIONAL BROADCASTERS	FREQUENCY
BBC World Service	6195
	7120
	7325
	9410
	9750
	9760
	12095
	15070
	17640
Christian Science Monitor World Service	7510
	9495
	9840
	9985
	13770
	17510
Radio Canada International	5995
	6050

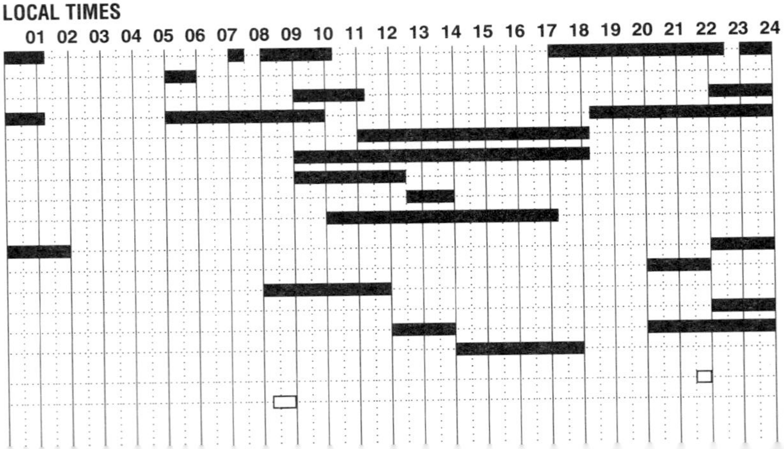

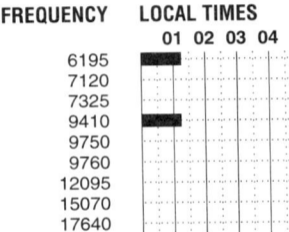

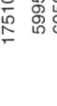

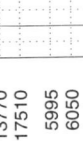

ATHENS

Frequency	Schedule
6150	
7235	
7295	
9750	
11775	
11945	23–24
15325	23–24
17840	
17875	
3965	20; 06–07
6045	03–04; 06–07
6175	06–07
7280	20
9745	03–04; 20
9805	16–17
11670	06–07
15195	13
15425	13

Radio France International

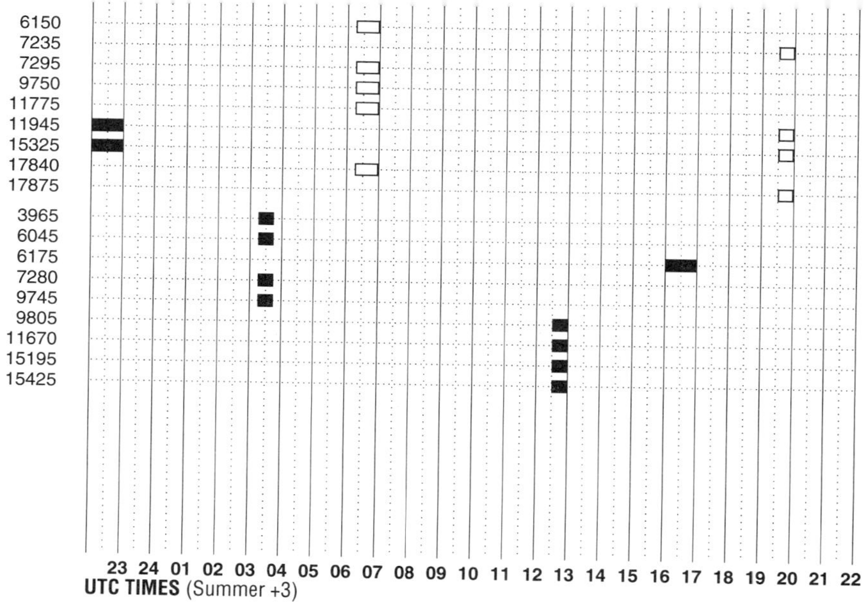

UTC TIMES (Summer +3)

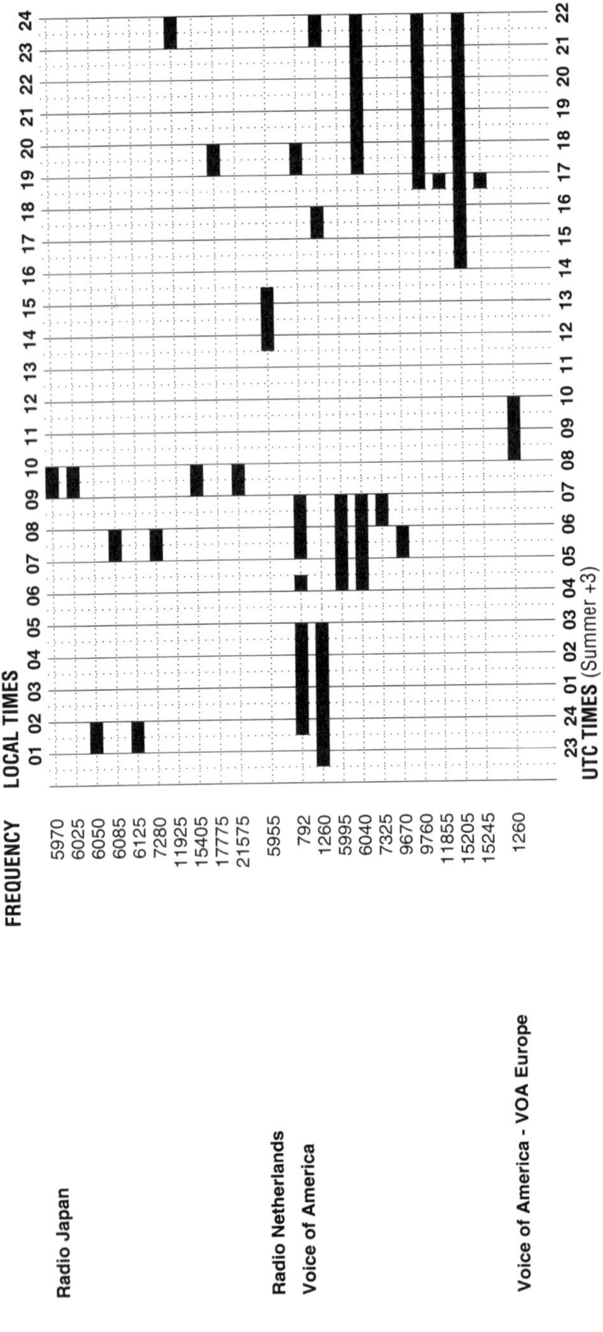

14 ATHENS

Station	Frequency
Radio Japan	5970, 6025, 6050, 6085, 6125, 7280, 11925, 15405, 17775, 21575
Radio Netherlands	5955
Voice of America	792, 1260, 5995, 6040, 7325, 9670, 9760, 11855, 15205, 15245
Voice of America - VOA Europe	1260

AUCKLAND 15

AUCKLAND, NEW ZEALAND
Hours difference from UTC (GMT): +12 (Summer +13)
Main business languages: English
Electricity: 50Hz, 230/400V
Currency: New Zealand Dollar
Telephone country prefix and area code: 64 (9)

LOCAL AM & FM STATIONS

	FREQUENCY
IXR - Arotearoa Radio Commercial station.	603
Radio Pacific Commercial station.	702
1YA - Radio National New Zealand Broadcasting Corporation national radio network. News on the hour.	756
IYC - Access Radio Radio New Zealand station. Broadcasts irregularly.	882
IZB - Newstalk 1ZB Radio New Zealand station.	1080
1XG - R.Rhema Commercial station.	1251

LOCAL TIMES
01 02 03 04 05 06 07 08 09 10 11 12 13 14 15 16 17 18 19 20 21 22 23 24

13 14 15 16 17 18 19 20 21 22 23 24 01 02 03 04 05 06 07 08 09 10 11 12
UTC TIMES (Summer +13)

16 AUCKLAND

FREQUENCY	LOCAL TIMES 01-24
1332	**1XI - Radio i** — Commercial station. News on the hour.
1476	**IXD Infomusic** — Commercial station.
FM 89.4	**1MMM - 89 FM** — Commercial station.
FM 90.2	**Fine Music** — Commercial station.
FM 91.0	**1MJK - 91 FM** — Commercial station.
FM 91.4	**1WCP - Concert Program** — Radio New Zealand station. Relays BBC News 0800, 1500, 2400.
FM 92.6	**1ACP - Concert Program** — Radio New Zealand station. Relays BBC News 0800, 1500, 2400.
FM 93.4	**2CBA - Kool 93FM** — Operated by Christian Broadcasting Association.
FM 95.0	**1STU B-FM** — Operated by Auckland University Students Association.

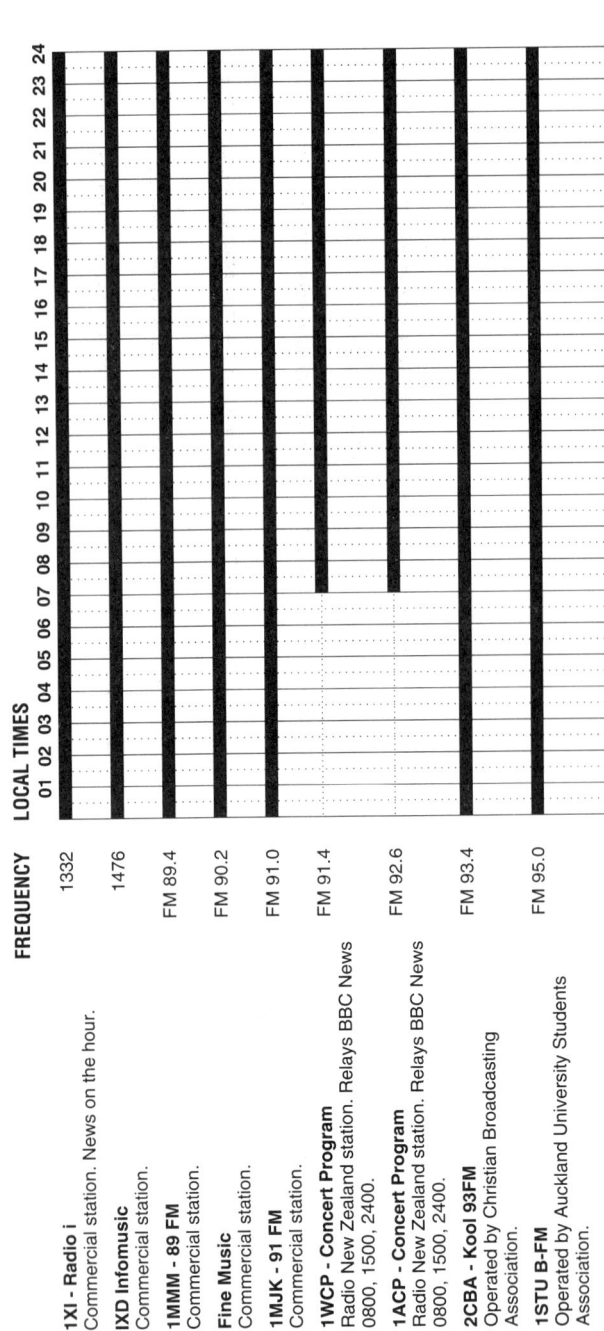

AUCKLAND 17

Radio i
Commercial station. News on the hour.

Hauraki 99FM
Commercial station. News on the hour.

FM 98.2

FM 99.0

INTERNATIONAL BROADCASTERS

BBC World Service

7150
9740
11955
15340
17830

Christian Science Monitor World Service

13615
13840
15665

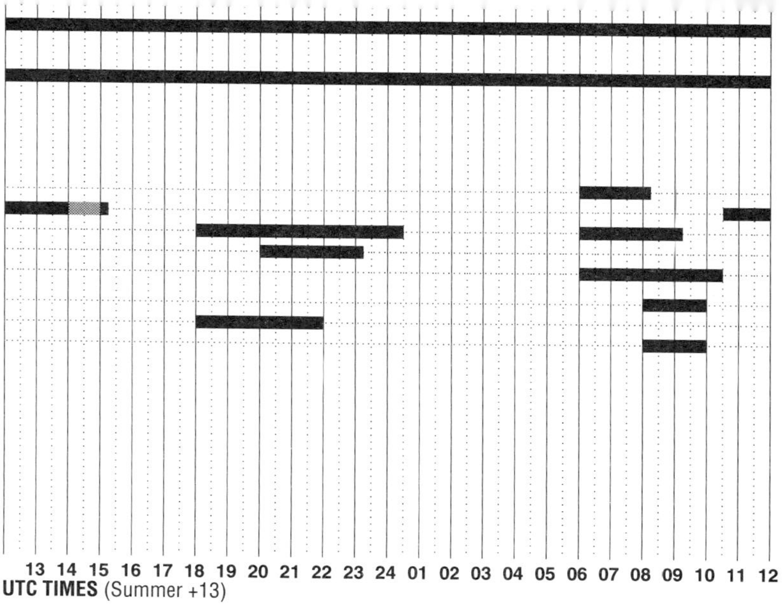

UTC TIMES (Summer +13)

18 AUCKLAND

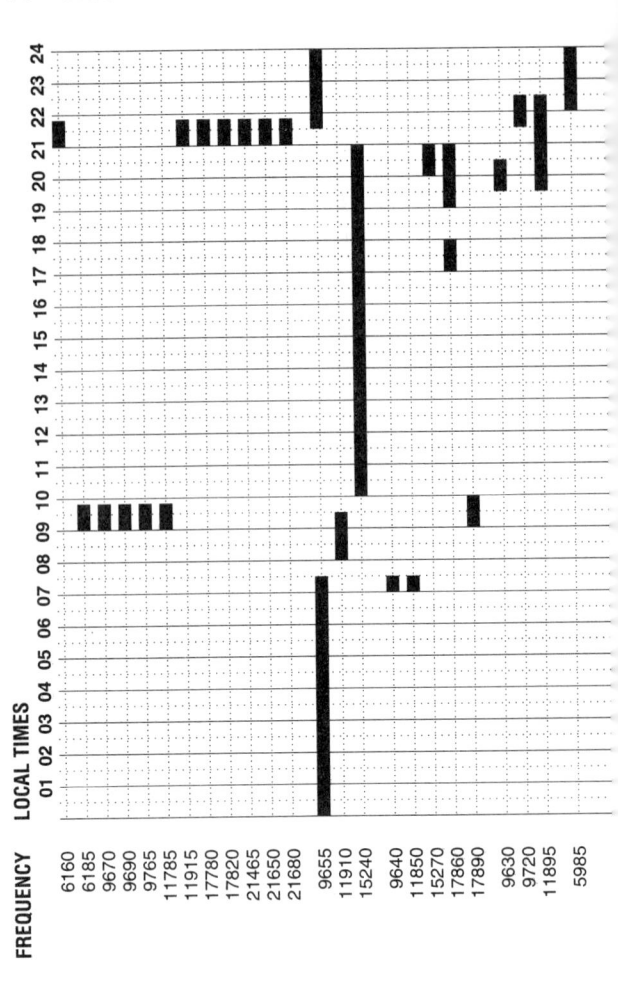

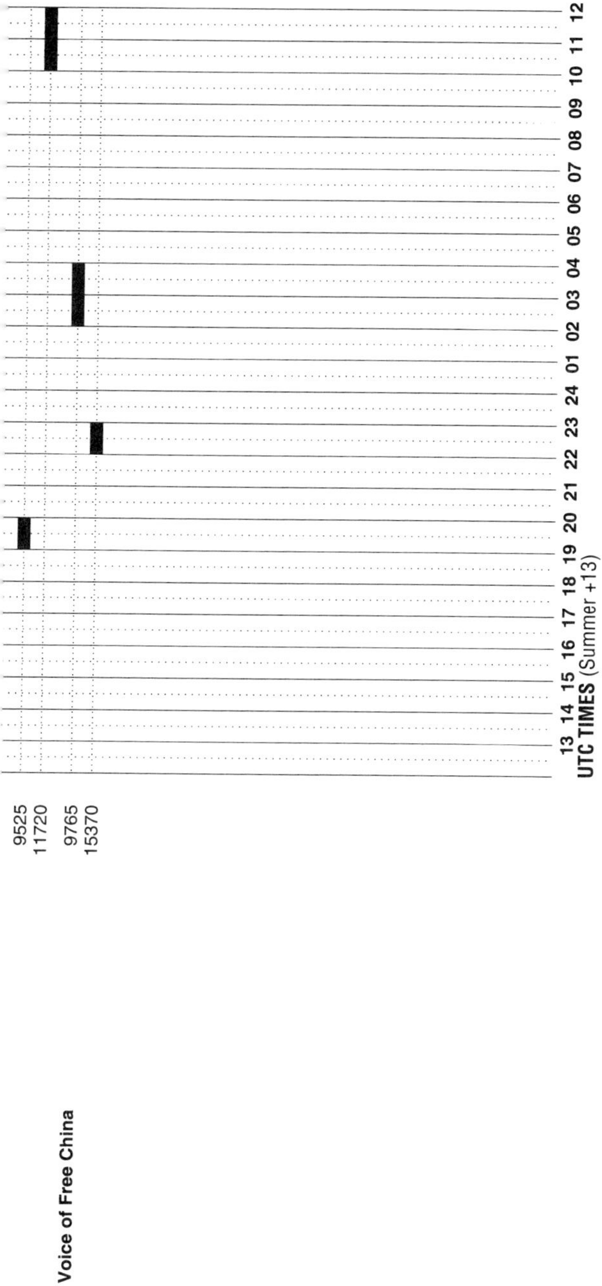

AUCKLAND 19

Voice of Free China

9525
11720
9765
15370

UTC TIMES (Summer +13)

20 BANGKOK

BANGKOK, THAILAND
Hours difference from UTC (GMT): +7
Main business languages: Thai, English
Electricity: 50Hz, 220/380V
Currency: Baht
Telephone country prefix and area code: 66 (2)

LOCAL AM & FM STATIONS

	FREQUENCY
Radio Thailand External Service Relay of program for overseas listeners.	927
95 FM Operated by Radio Thailand.	FM 95.5
Radio Thailand External Service Relay of program for overseas listeners.	FM 97.0

INTERNATIONAL BROADCASTERS

BBC World Service
7180
9740
11820
11955
15280
15360
17830
21715

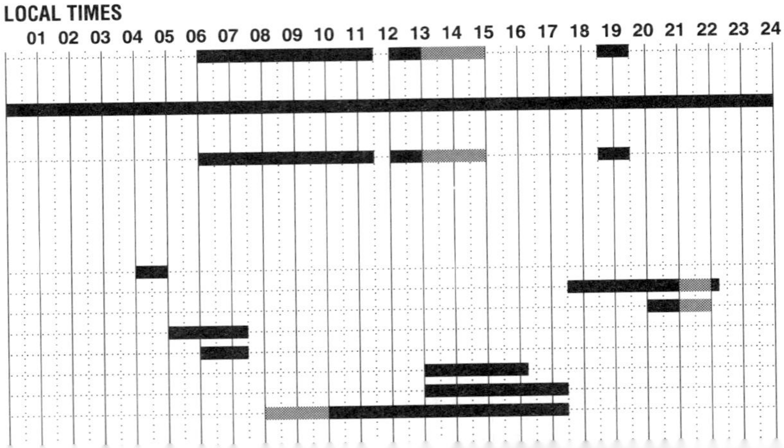

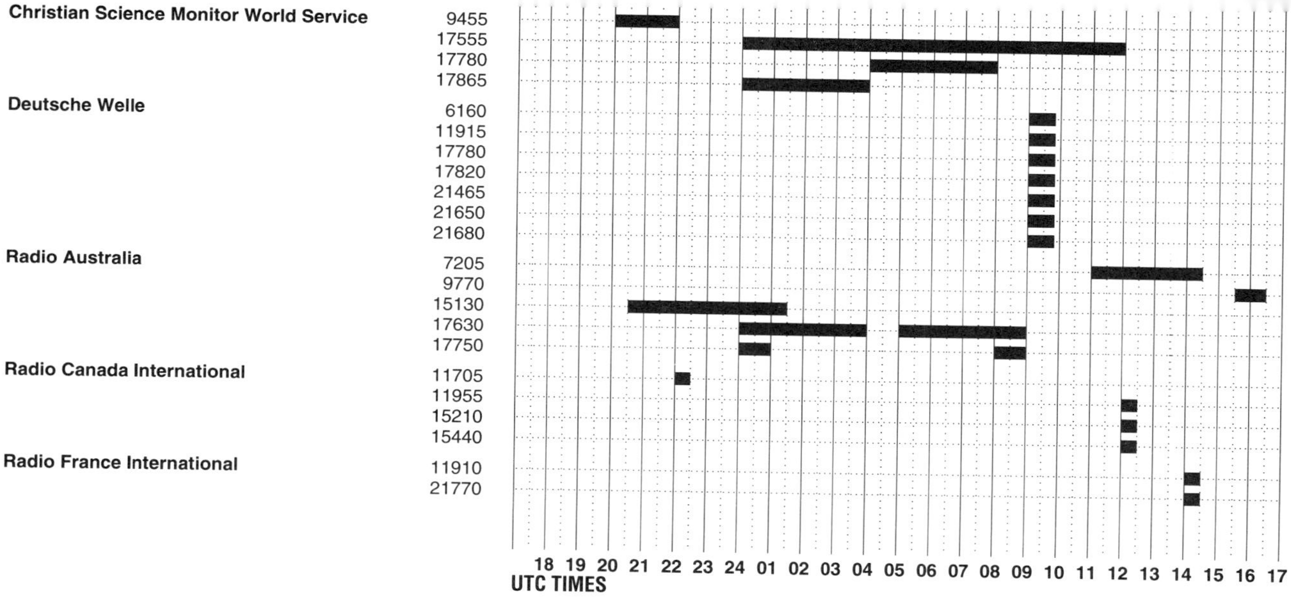

BANGKOK 21

Station	Frequencies
Christian Science Monitor World Service	9455, 17555, 17780, 17865
Deutsche Welle	6160, 11915, 17780, 17820, 21465, 21650, 21680
Radio Australia	7205, 9770, 15130, 17630, 17750
Radio Canada International	11705, 11955, 15210, 15440
Radio France International	11910, 21770

22 BANGKOK

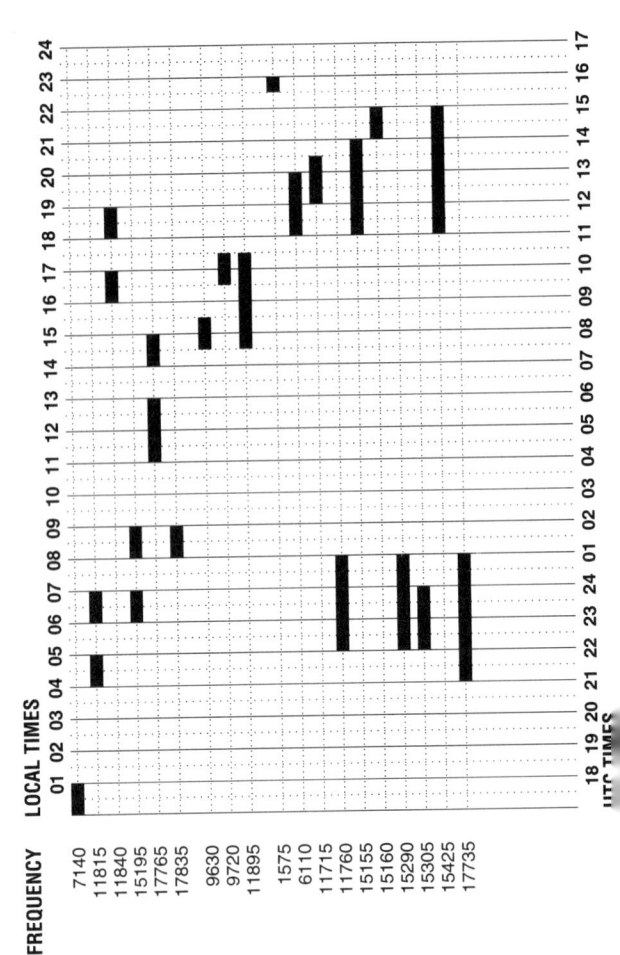

Station	Frequency	Schedule
Radio Japan	7140	
	11815	
	11840	
	15195	
	17765	
	17835	
Radio Netherlands	9630	
	9720	
	11895	
Voice of America	1575	
	6110	
	11715	
	11760	
	15155	
	15160	
	15290	
	15305	
	15425	
	17735	

BARCELONA, SPAIN

Hours difference from UTC (GMT): +1 (Summer +2)
Main business languages: Spanish, English
Electricity: 50Hz, voltage varies
Currency: Peseta
Telephone country prefix and area code: 34 (3)

INTERNATIONAL BROADCASTERS	FREQUENCY
BBC World Service	3955
	6195
	7120
	7325
	9410
	9750
	9760
	12095
	15070
	15575
	17640
	17705

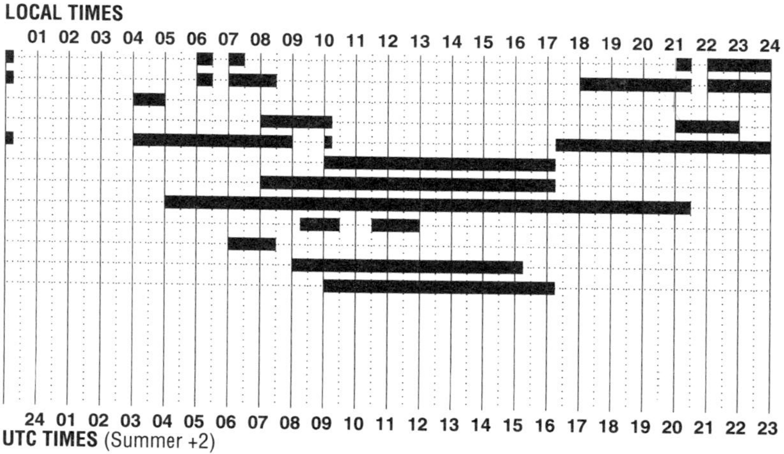

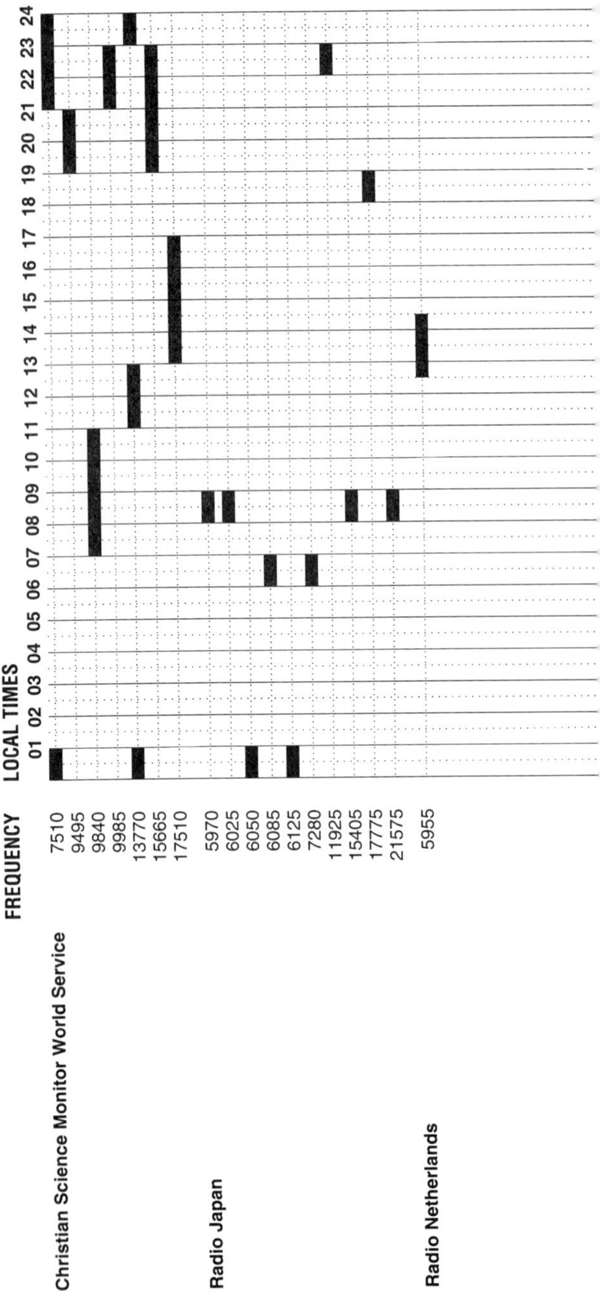

24 BARCELONA

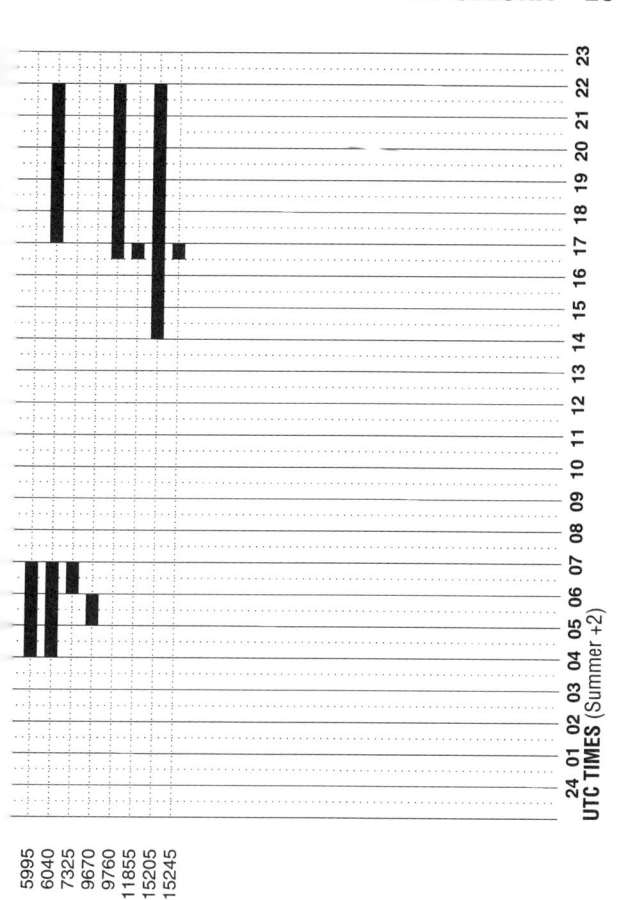

BARCELONA 25

Voice of America

5995
6040
7325
9670
11855
15205
15245

UTC TIMES (Summer +2)

26 BEIJING

BEIJING, CHINA
Hours difference from UTC (GMT): +8
Main business languages: Chinese, English
Electricity: 50Hz, 220V
Currency: Yuan
Telephone country prefix and area code: 86 (1)

LOCAL AM & FM STATIONS

	FREQUENCY
Radio Beijing Capital Service	1251
Local news, travel, features and relays of External Service.	FM 91.5

INTERNATIONAL BROADCASTERS

BBC World Service

7180
9570
9740
11765
11820
11945
11955
15280
15360
17790
17830
21715

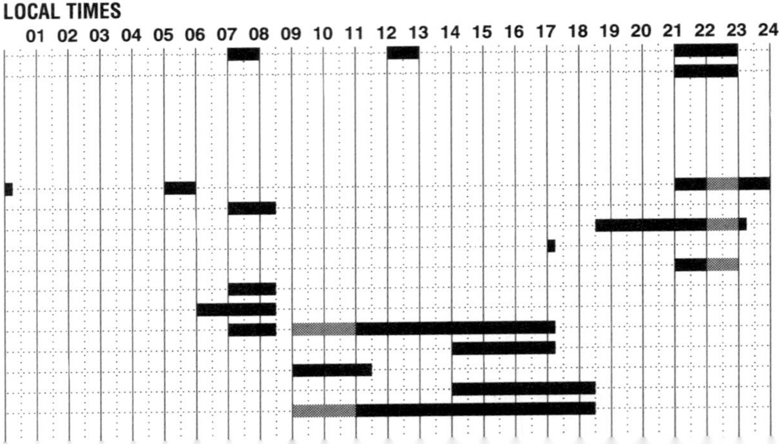

LOCAL TIMES
01 02 03 04 05 06 07 08 09 10 11 12 13 14 15 16 17 18 19 20 21 22 23 24

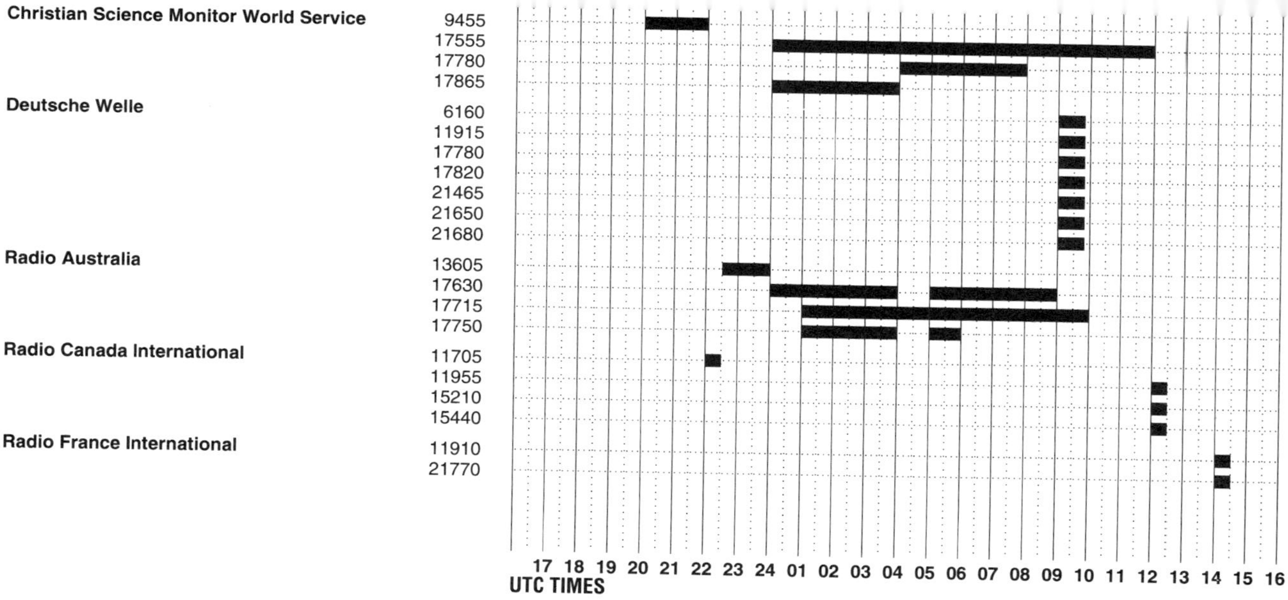

BEIJING

Station	Frequency
Christian Science Monitor World Service	9455
	17555
	17780
	17865
Deutsche Welle	6160
	11915
	17780
	17820
	21465
	21650
	21680
Radio Australia	13605
	17630
	17715
	17750
Radio Canada International	11705
	11955
	15210
	15440
Radio France International	11910
	21770

UTC TIMES

28 BEIJING

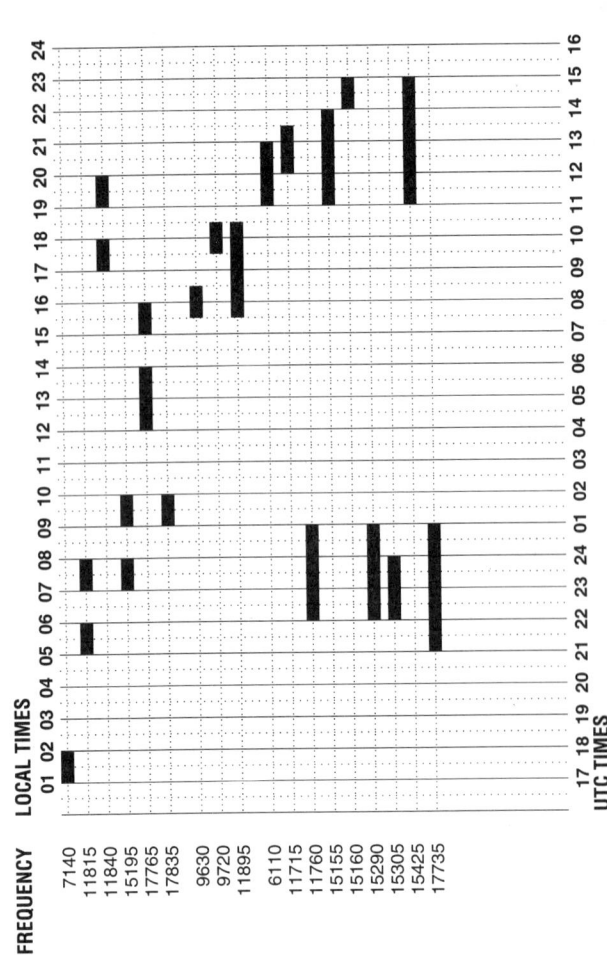

BERLIN, GERMANY

Hours difference from UTC (GMT): +1 (Summer +2)
Main business languages: German, English
Electricity: 50Hz, 220V
Currency: Deutschmark
Telephone country prefix and area code: 49 (30)

LOCAL AM & FM STATIONS

	FREQUENCY
American Forces Network News on the hour. Includes some local programs.	1107
American Forces Network 2 Easy listening music station.	FM 87.9
BBC World Service Same programs as shortwave, with satellite audio quality.	FM 90.2
British Forces Broadcasting Service Local programs and feeds from London. News on the hour.	FM 98.8
VOA Europe Operated by the Voice of America. Available on hotel cable system only. Alternate to 91.0	FM 106.3

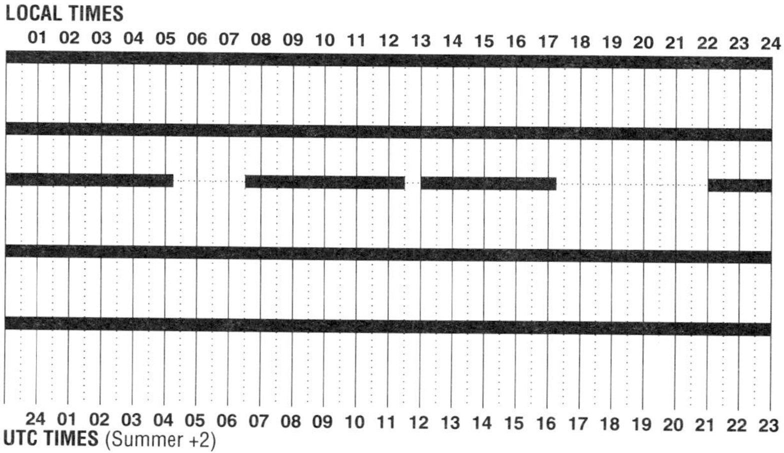

BERLIN 29

30 BERLIN

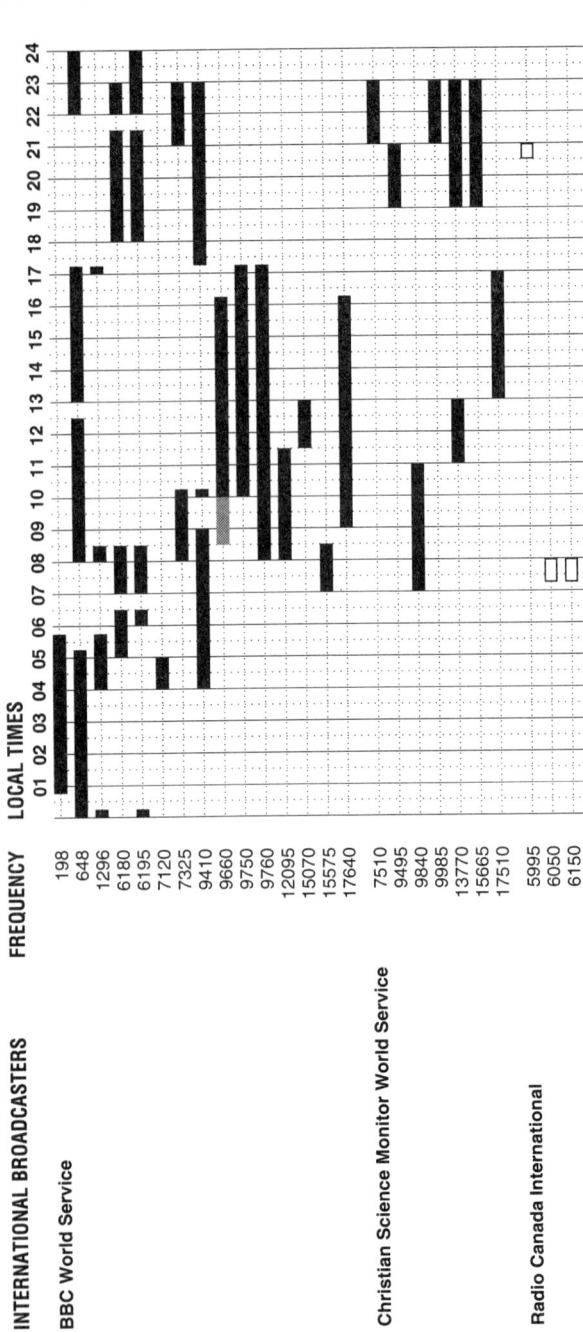

INTERNATIONAL BROADCASTERS	FREQUENCY
BBC World Service	198
	648
	1296
	6180
	6195
	7120
	7325
	9410
	9660
	9750
	9760
	12095
	15070
	15575
	17640
Christian Science Monitor World Service	7510
	9495
	9840
	9985
	13770
	15665
	17510
Radio Canada International	5995
	6050
	6150

BERLIN 31

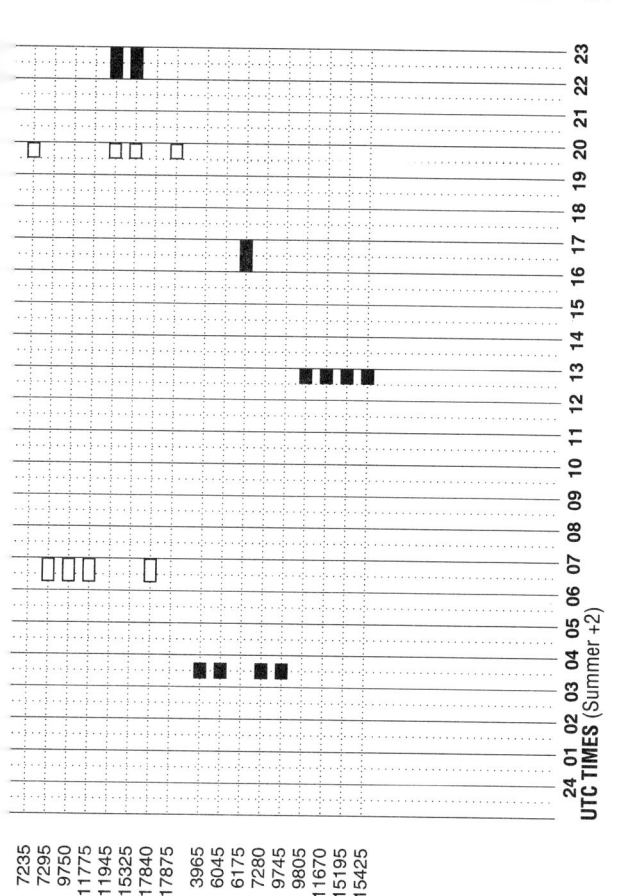

Radio France International

Frequencies
7235
7295
9750
11775
11945
15325
17840
17875
3965
6045
6175
7280
9745
9805
11670
15195
15425

UTC TIMES (Summer +2)

32 BERLIN

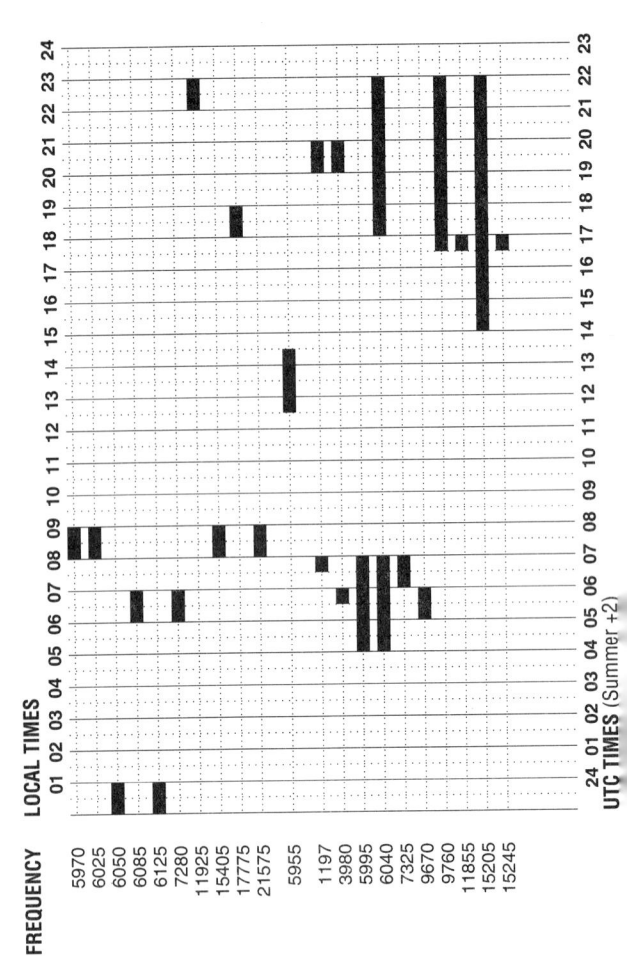

Station	Frequency
Radio Japan	5970, 6025, 6050, 6085, 6125, 7280, 11925, 15405, 17775, 21575
Radio Netherlands	5955
Voice of America	1197, 3980, 5995, 6040, 7325, 9670, 9760, 11855, 15205, 15245

BOMBAY 33

BOMBAY, INDIA
Hours difference from UTC (GMT): +5.5
Main business languages: English
Electricity: 50Hz, 220/380V
Currency: Indian Rupee
Telephone country prefix and area code: 91 (22)

LOCAL AM & FM STATIONS	FREQUENCY
All India Radio Program B	558
All India Radio Program A	1044
All India Radio Program C	1188
All India Radio FM Program	FM 107.1

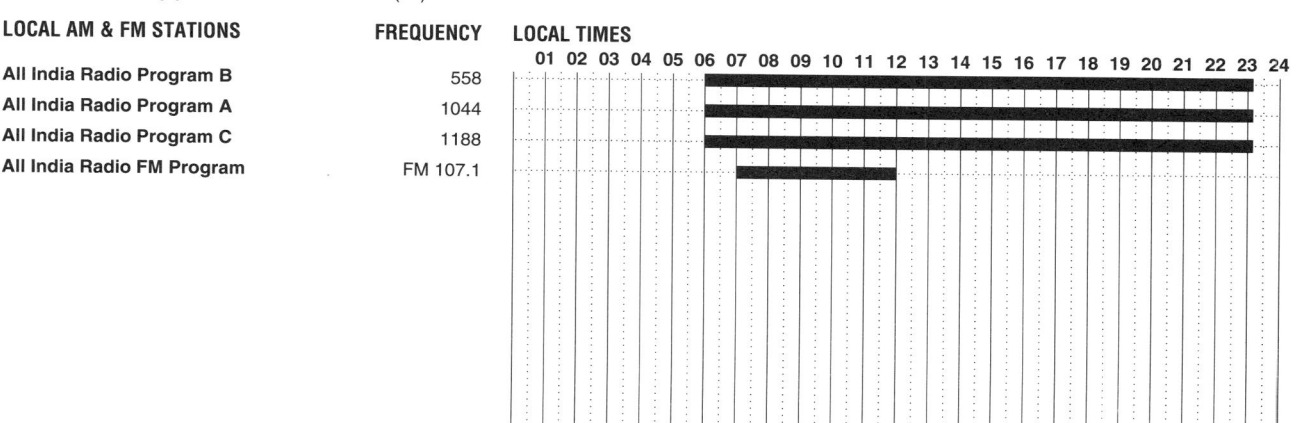

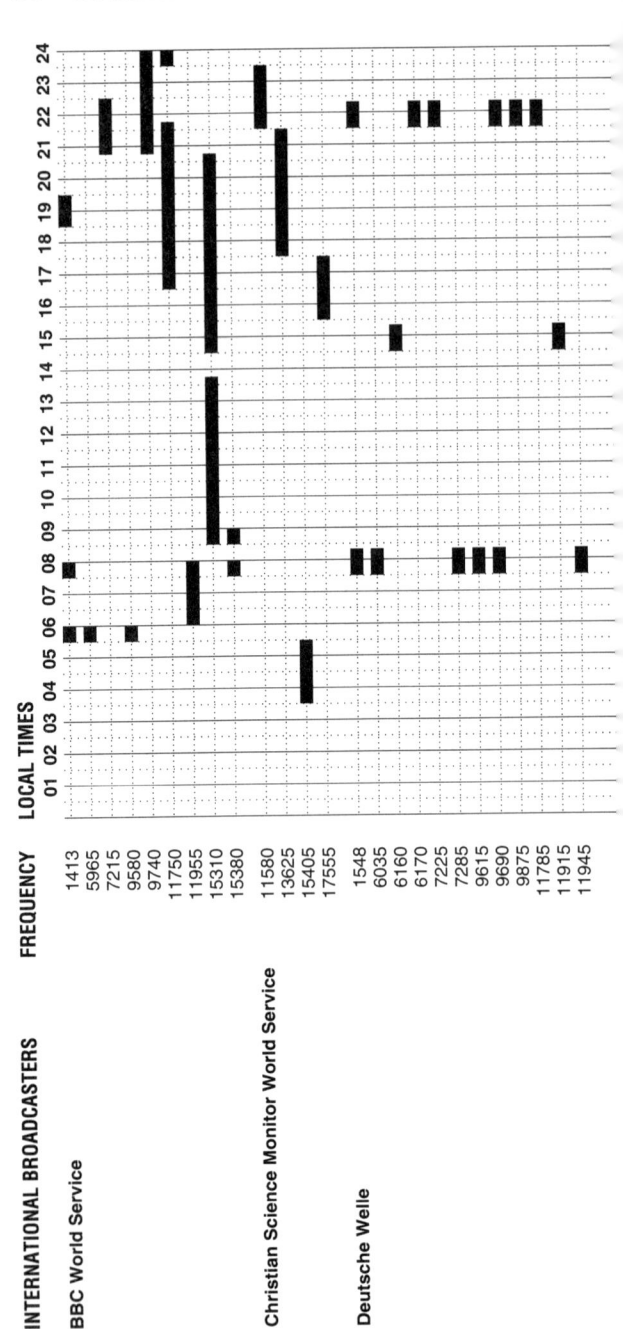

34 BOMBAY

INTERNATIONAL BROADCASTERS	FREQUENCY
BBC World Service	1413
	5965
	7215
	9580
	9740
	11750
	11955
	15310
	15380
Christian Science Monitor World Service	11580
	13625
	15405
	17555
Deutsche Welle	1548
	6035
	6160
	6170
	7225
	7285
	9615
	9690
	9875
	11785
	11915
	11945

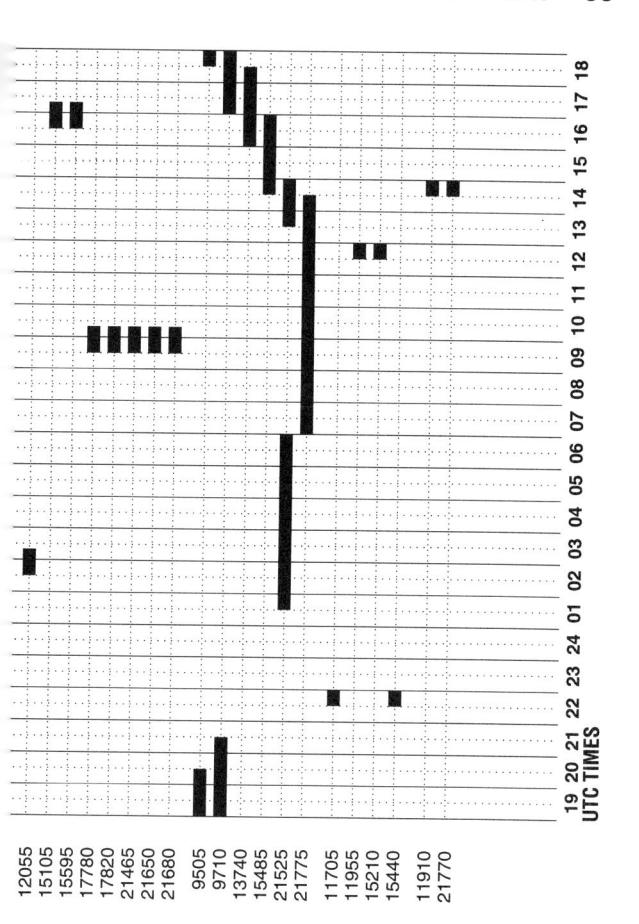

BOMBAY 35

Frequency	Station
12055	
15105	
15595	
17780	
17820	
21465	
21650	
21680	Radio Australia
9505	
9710	
13740	
15485	
21525	
21775	
11705	
11955	
15210	
15440	Radio Canada International
11910	
21770	Radio France International

UTC TIMES

36 BOMBAY

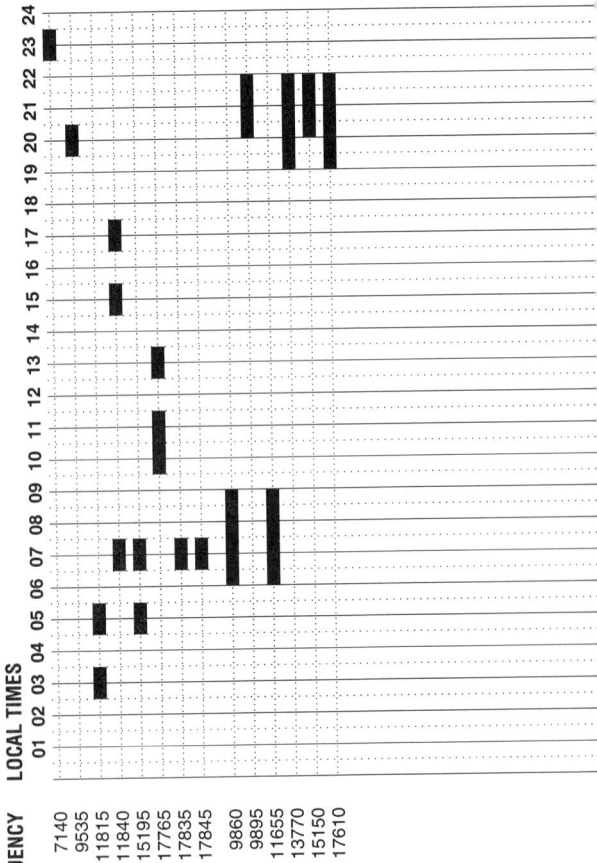

Frequency	Station	Local Times
7140	Radio Japan	23–24
9535	Radio Japan	19–20
11815	Radio Japan	16–17
11840	Radio Japan	04–05
15195	Radio Japan	14–15, 05–06
17765	Radio Japan	13–14
17835	Radio Japan	10–12
17845	Radio Japan	02–03
9860	Radio Netherlands	06–07
9895	Radio Netherlands	19–22
11655	Radio Netherlands	06–07
13770	Radio Netherlands	20–22
15150	Radio Netherlands	06–09
17610	Radio Netherlands	06–09, 20–22

BOMBAY 37

Voice of America

Frequencies (kHz): 6070, 6110, 7120, 9560, 9645, 9760, 9770, 11705, 11855, 15185, 15425, 17740, 21550

UTC TIMES

38 BRUSSELS

BRUSSELS, BELGIUM
Hours difference from UTC (GMT): +1 (Summer +2)
Main business languages: French, English
Electricity: 50Hz, 220V
Currency: Belgian Francs
Telephone country prefix and area code: 32 (2)

LOCAL AM & FM STATIONS

Station	Frequency
BBC Radio 4 British domestic network. Main news 0800, 0900, 1400, 1800, 1900, 2300, 0100.	198
BBC Radio 5 British sports and educational network.	693 909
BRT Brussels Calling Relay of program for listeners abroad.	1512
American Forces Network News on the hour by satellite from the U.S.	FM 101.7

INTERNATIONAL BROADCASTERS

BBC World Service	198 648 1296 6180

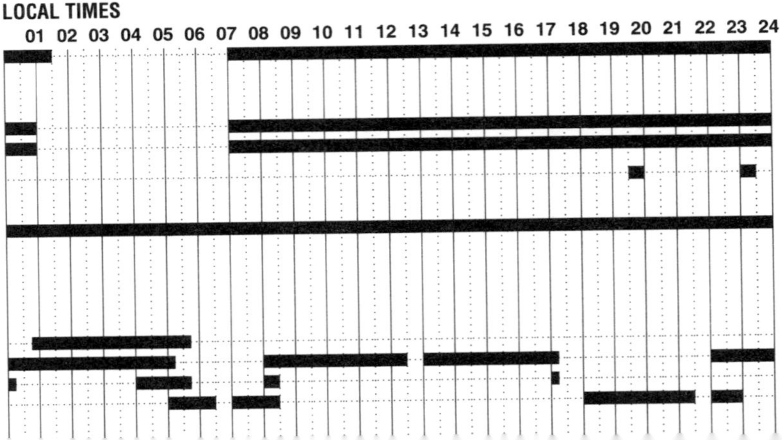

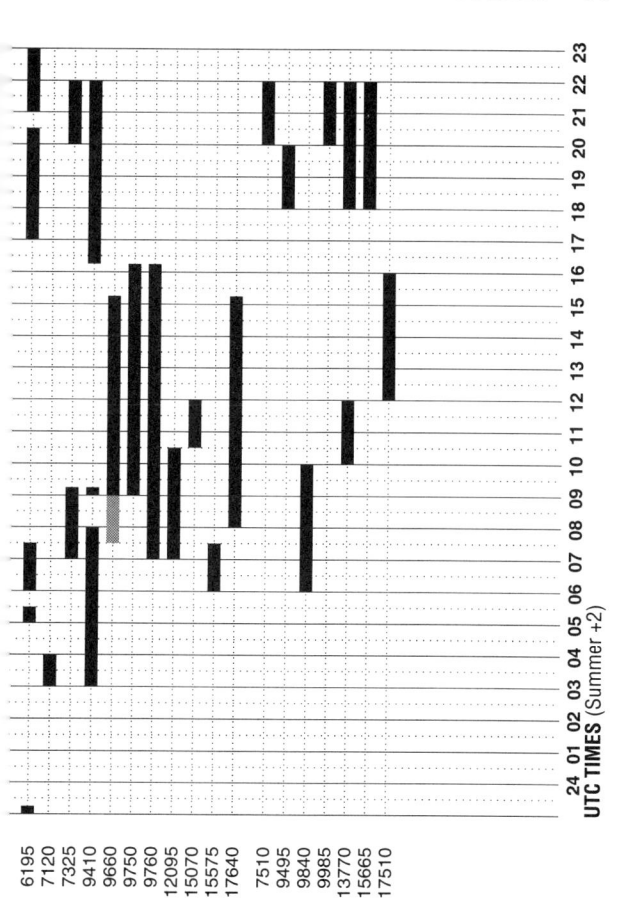

BRUSSELS 39

Christian Science Monitor World Service

Frequency
6195
7120
7325
9410
9660
9750
9760
12095
15070
15575
17640
7510
9495
9840
9985
13770
15665
17510

UTC TIMES (Summer +2)

40 BRUSSELS

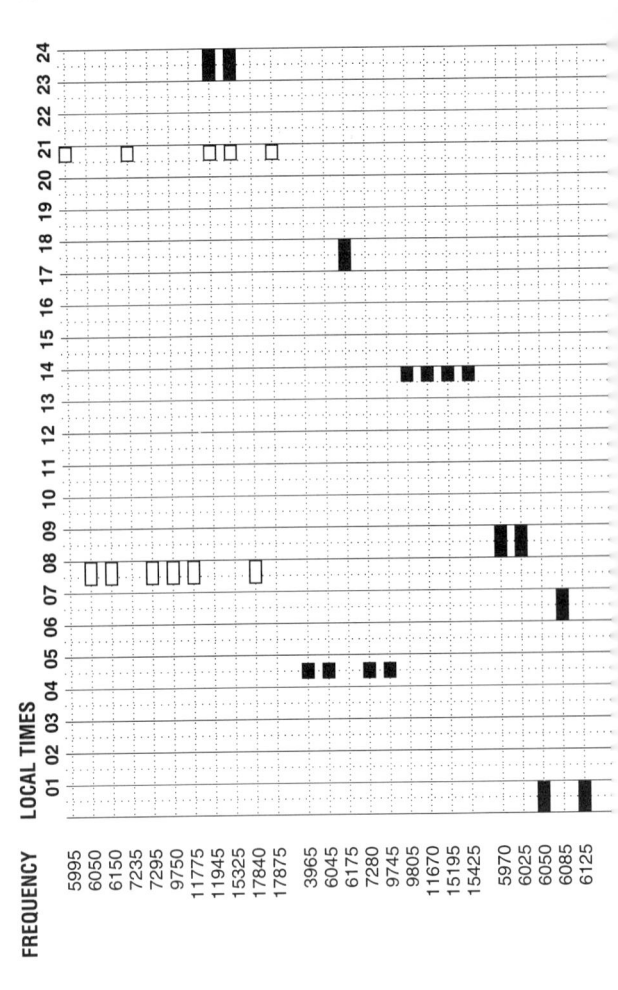

Radio Canada International

Frequency
5995
6050
6150
7235
7295
9750
11775
11945
15325
17840
17875

Radio France International

Frequency
3965
6045
6175
7280
9745
9805
11670
15195
15425

Radio Japan

Frequency
5970
6025
6050
6085
6125

BRUSSELS 41

Radio Netherlands
7280
11925
15405
17775
21575

Voice of America
5955
1197
3980
5995
6040
7325
9670
11855
15205
15245

Voice of America - VOA Europe
1197

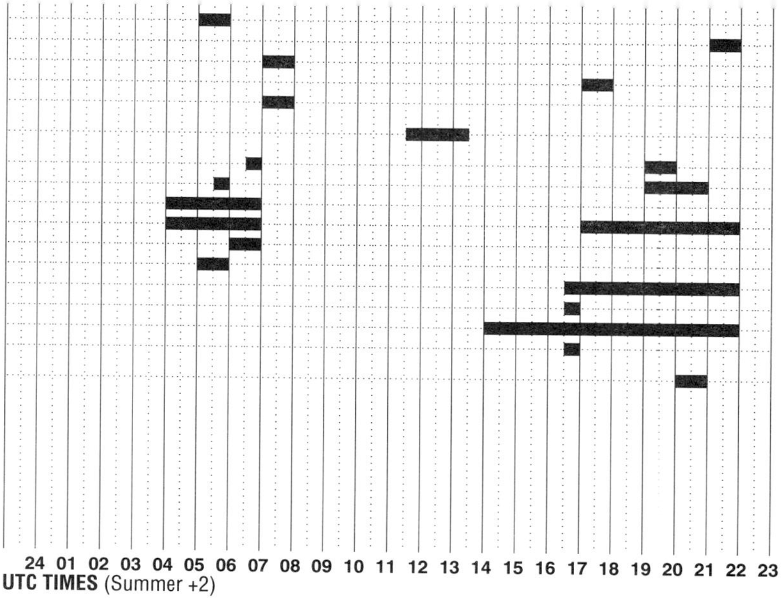

UTC TIMES (Summer +2)

42 BUENOS AIRES

BUENOS AIRES, ARGENTINA
Hours difference from UTC (GMT): -3 (Summer -2)
Main business languages: Spanish, English
Electricity: 50Hz, 220V
Currency: Austral
Telephone country prefix and area code: 54 (1)

INTERNATIONAL BROADCASTERS	FREQUENCY
BBC World Service	6005
	9915
	11750
	15190
	15220
	15260
	17840
Christian Science Monitor World Service	9455
	9870
	13760
	17555
HCJB, Ecuador	15115
Radio Canada International	9535
	11845
	11940
	13720

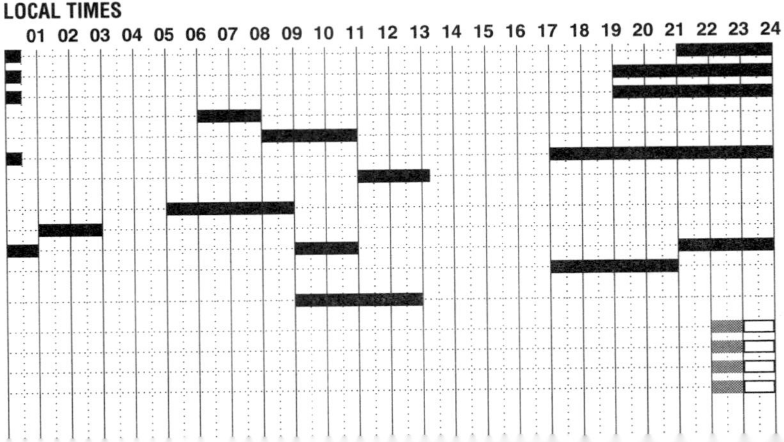

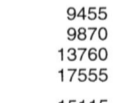

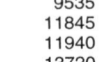

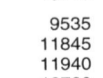

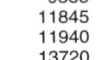

BUENOS AIRES 43

Radio France International 21810

Radio Japan 17825, 21610

Voice of America 5995, 6130, 7405, 9455, 9590, 9775, 11580, 11695, 11915, 15120, 15205

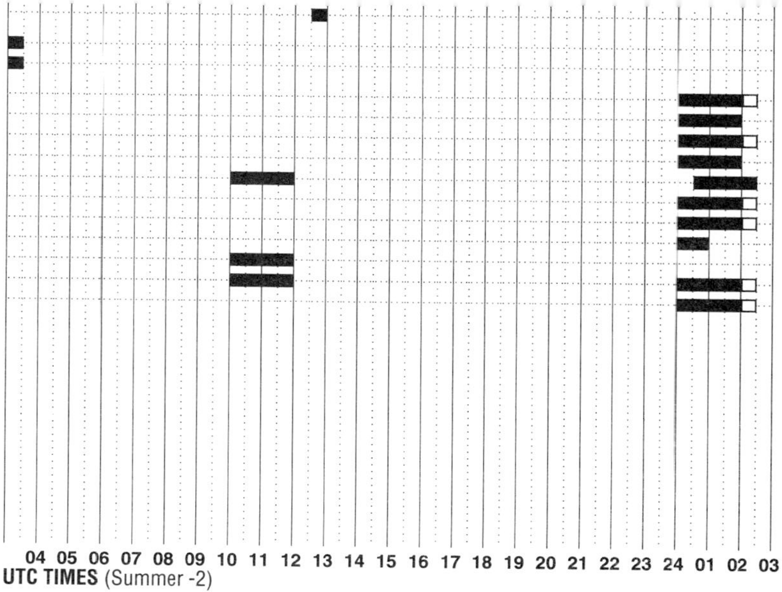

UTC TIMES (Summer -2)

44　CAIRO

CAIRO, EGYPT
Hours difference from UTC (GMT): +2 (Summer +3)
Main business languages: Arabic, English
Electricity: 50Hz, 220V
Currency: Egyptian Pound
Telephone country prefix and area code: 20 (2)

LOCAL AM & FM STATIONS | FREQUENCY

Radio Cairo European Program | 558
Also 0900-1200 on Friday and Saturday. | FM 95.4

INTERNATIONAL BROADCASTERS

BBC World Service
 639
 702
 720
 1323
 1413
 5965
 6195
 9580
 11760
 11955

Christian Science Monitor World Service
 9495
 13770

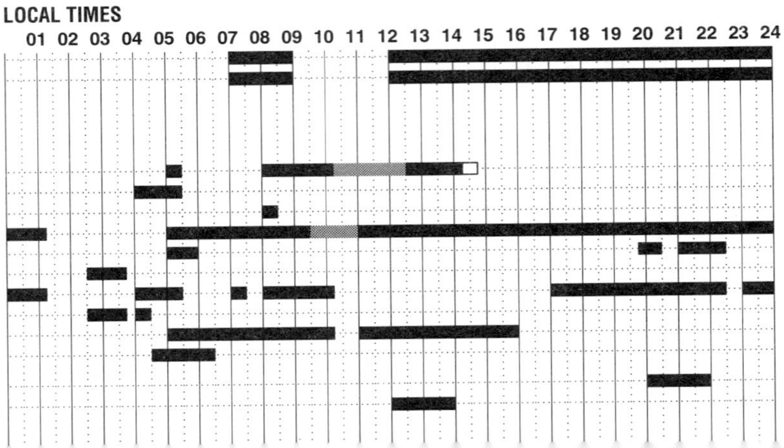

CAIRO 45

Radio Canada International

17510
21640

Radio Japan

15275

5970
6025
6050
6085
6125
7280
11925
15405
17775
21575

UTC TIMES (Summer +3)

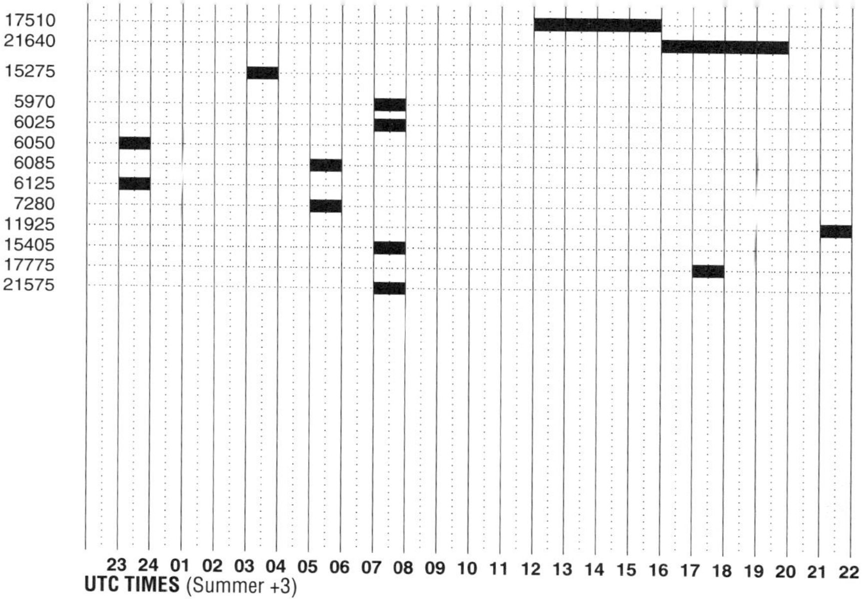

46 CAIRO

FREQUENCY	
6060	
7205	
9530	
9700	
9740	
11735	
11825	
11905	
11960	
15160	
15205	
15445	
21455	
21570	

Voice of America

FREQUENCY	
11735	
15160	
15195	
21455	
21570	

Voice of America - VOA Europe

LOCAL TIMES
UTC TIMES (Summer +3)

CALCUTTA 47

CALCUTTA, INDIA
Hours difference from UTC (GMT): +5.5
Main business languages: English
Electricity: 50Hz, 220/380V
Currency: Indian Rupee
Telephone country prefix and area code: 91 (33)

LOCAL AM & FM STATIONS	FREQUENCY
All India Radio Program A	657
All India Radio Program B	1008
All India Radio Program D	1224
All India Radio Program C	1323
All India Radio FM Program	FM 107.1

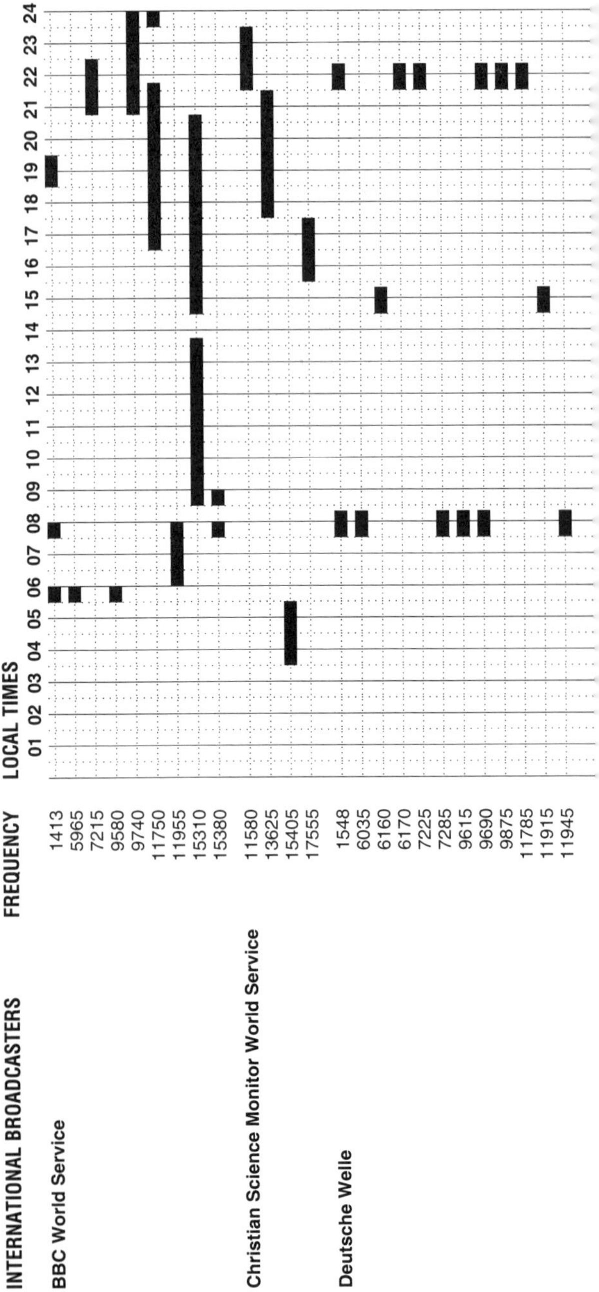

48 CALCUTTA

INTERNATIONAL BROADCASTERS	FREQUENCY
BBC World Service	1413
	5965
	7215
	9580
	9740
	11750
	11955
	15310
	15380
Christian Science Monitor World Service	11580
	13625
	15405
	17555
Deutsche Welle	1548
	6035
	6160
	6170
	7225
	7285
	9615
	9690
	9875
	11785
	11915
	11945

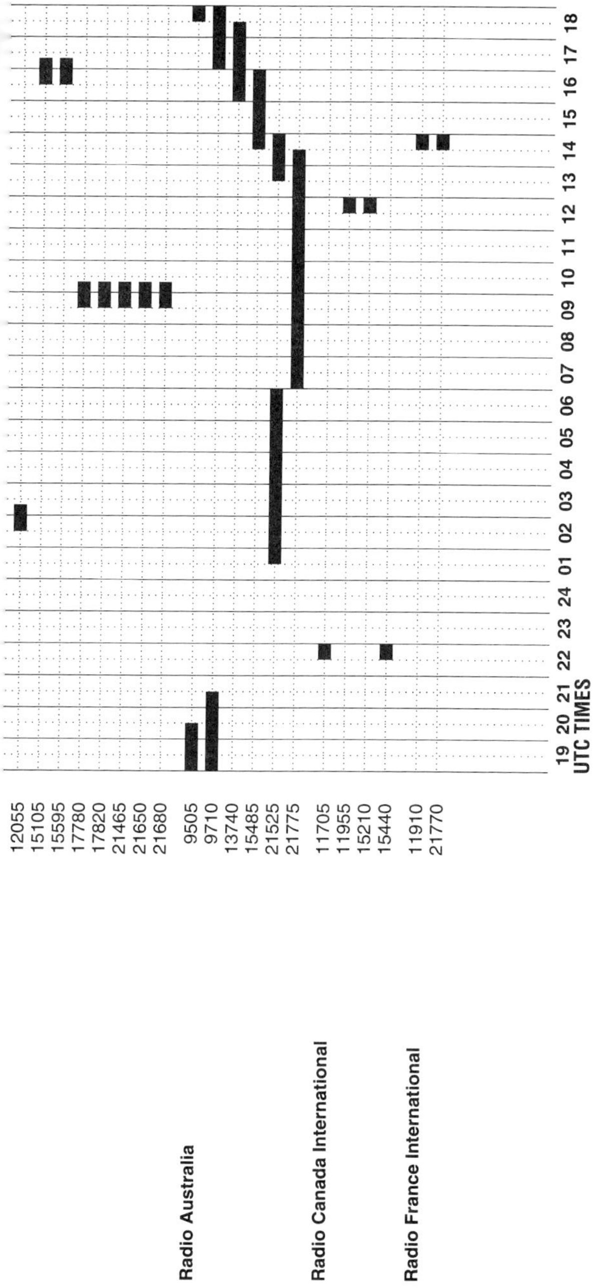

CALCUTTA 49

	12055
15105
15595
17780
21465
21650
21680
Radio Australia	9505
9710
13740
15485
21525
21775
Radio Canada International	11705
11955
15210
15440
Radio France International	11910
21770

UTC TIMES

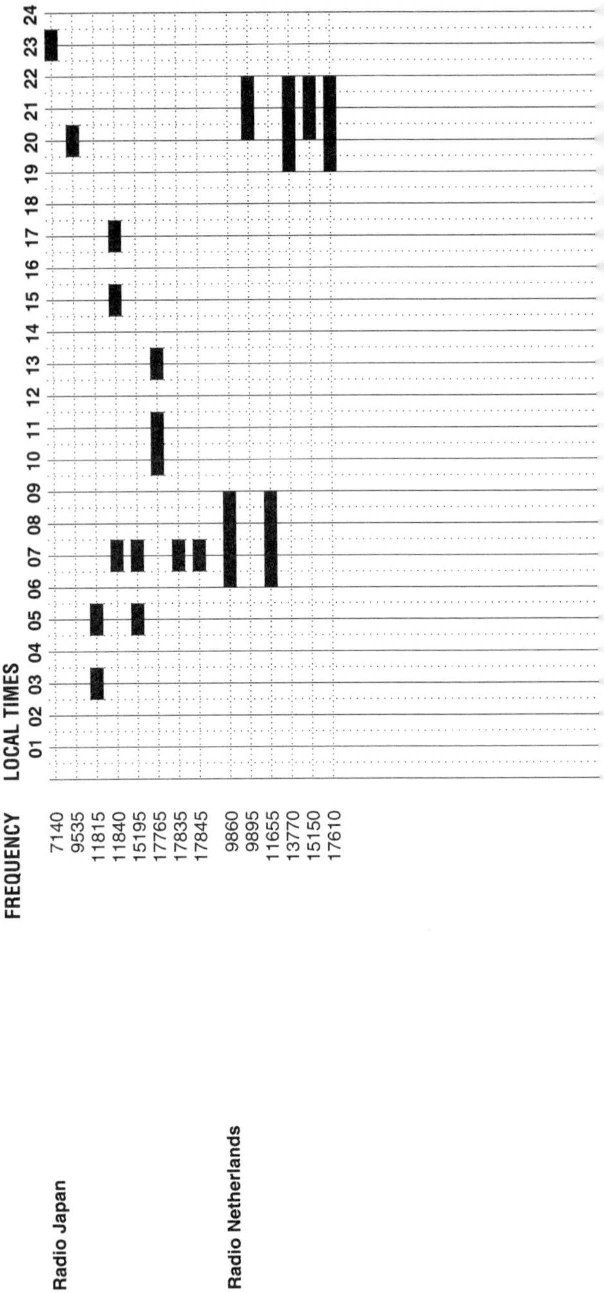

50 CALCUTTA

FREQUENCY

Radio Japan
- 7140
- 9535
- 11815
- 11840
- 15195
- 17765
- 17835
- 17845

Radio Netherlands
- 9860
- 9895
- 11655
- 13770
- 15150
- 17610

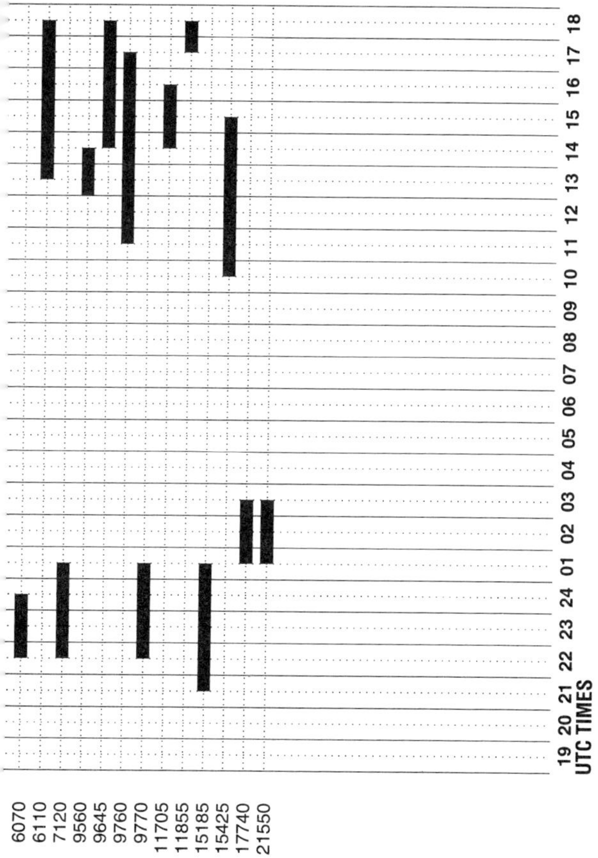

CALCUTTA 51

Voice of America

52 CARACAS

CARACAS, VENEZUELA
Hours difference from UTC (GMT): -4
Main business languages: Spanish, English
Electricity: 60Hz, 120/240V
Currency: Bolivar
Telephone country prefix and area code: 58 (2)

LOCAL AM & FM STATIONS

	FREQUENCY
Trans World Radio Bonaire Religious station. News 0700.	800

INTERNATIONAL BROADCASTERS

BBC World Service	5975 6005 6195 9590 9640 9915 11750 15190 15220 15260 17840
Christian Science Monitor World Service	9455

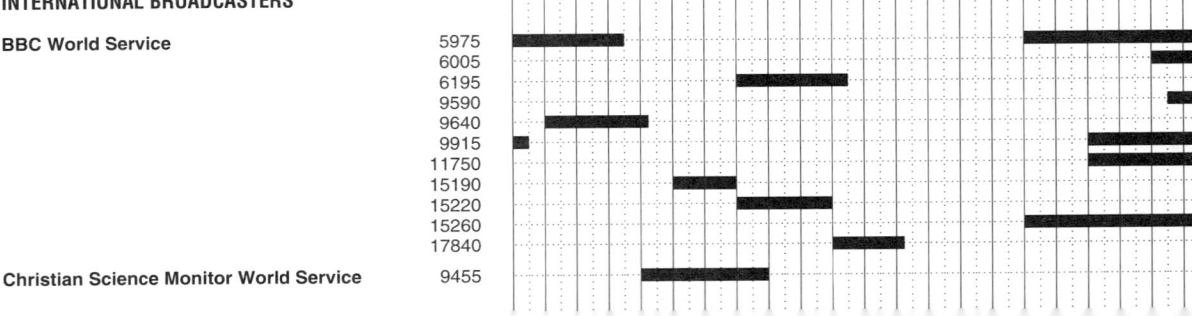

CARACAS 53

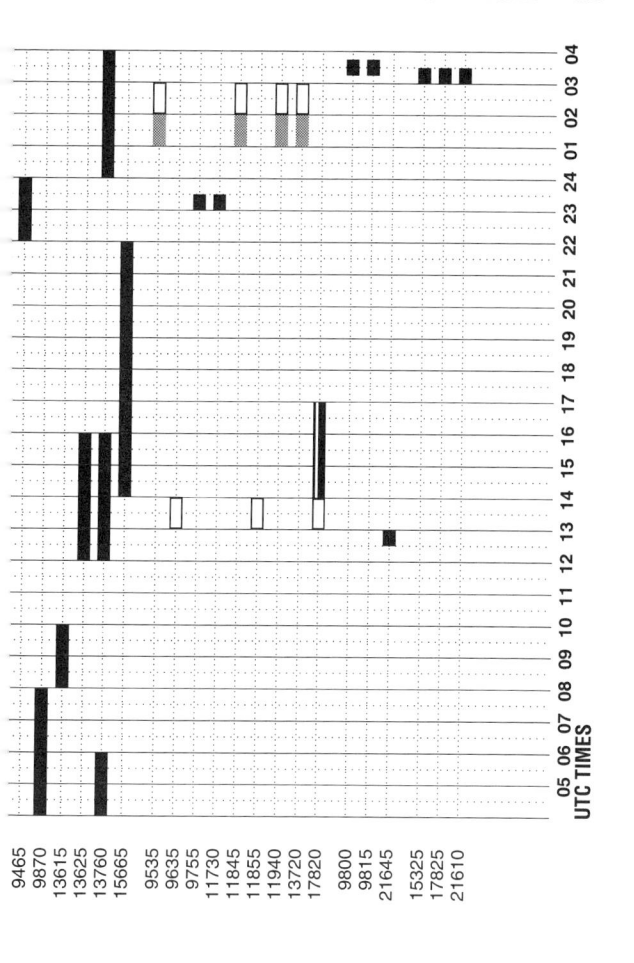

54 CARACAS

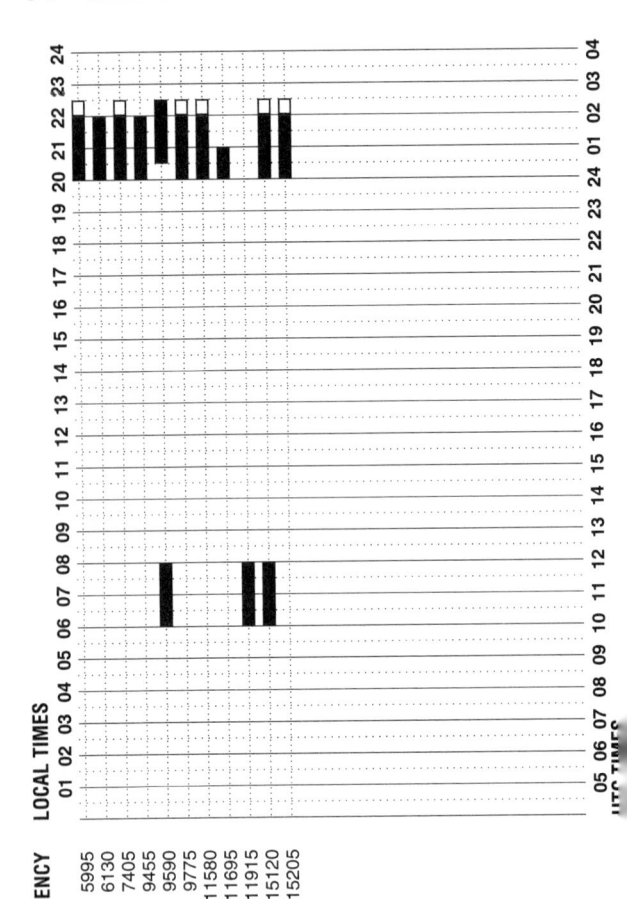

Voice of America

FREQUENCY	Broadcast times
5995	
6130	
7405	
9455	
9590	
9775	
11580	
11695	
11915	
15120	
15205	

CHICAGO 55

Hours difference from UTC (GMT): -6 (Summer -5)
Main business languages: English
Electricity: 60Hz, 110V
Currency: U.S.Dollar
Telephone country prefix and area code: 1 (312)

LOCAL AM & FM STATIONS

	FREQUENCY
WMAQ All news.	670
WGN Talk.	720
WBBM All news.	780
WLS Talk.	890
WBEZ News, talk, jazz (non-commercial station).	FM 91.5
WNIB Fine arts.	FM 97.1
WFMT Fine arts.	FM 98.7

LOCAL TIMES
01 02 03 04 05 06 07 08 09 10 11 12 13 14 15 16 17 18 19 20 21 22 23 24

UTC TIMES (Summer -5)
07 08 09 10 11 12 13 14 15 16 17 18 19 20 21 22 23 24 01 02 03 04 05 06

56 CHICAGO

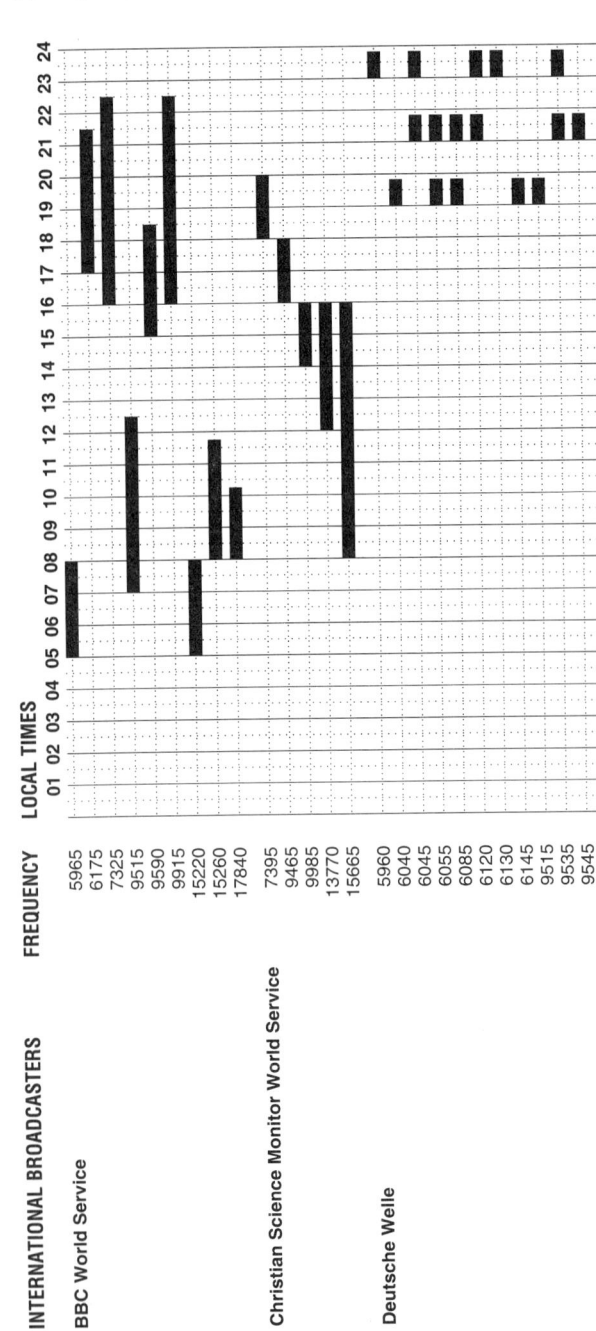

INTERNATIONAL BROADCASTERS	FREQUENCY
BBC World Service	5965
	6175
	7325
	9515
	9590
	9915
	15220
	15260
	17840
Christian Science Monitor World Service	7395
	9465
	9985
	13770
	15665
Deutsche Welle	5960
	6040
	6045
	6055
	6085
	6120
	6130
	6145
	9515
	9535
	9545

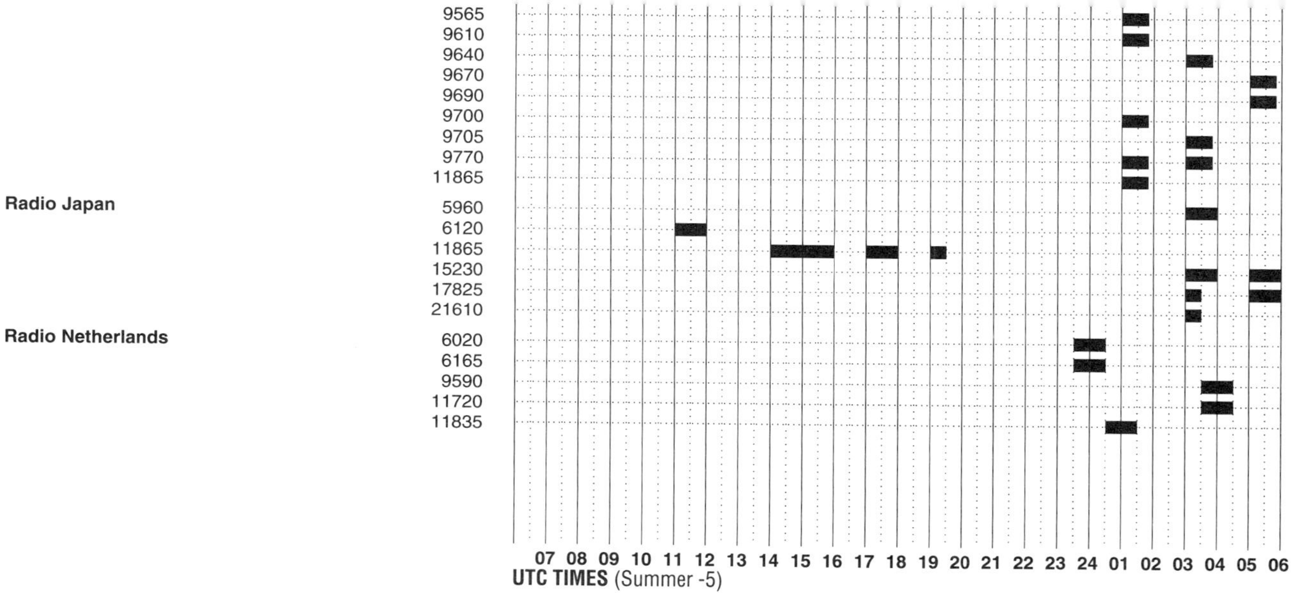

CHICAGO 57

Radio Japan

Frequency	
9565	
9610	
9640	
9670	
9690	
9700	
9705	
9770	
11865	
5960	
6120	
11865	
15230	
17825	
21610	

Radio Netherlands

Frequency	
6020	
6165	
9590	
11720	
11835	

UTC TIMES (Summer −5)
07 08 09 10 11 12 13 14 15 16 17 18 19 20 21 22 23 24 01 02 03 04 05 06

58 COPENHAGEN

COPENHAGEN, DENMARK
Hours difference from UTC (GMT): +1 (Summer +2)
Main business languages: Danish, English
Electricity: 50Hz, 220V
Currency: Danish Krone
Telephone country prefix and area code: 45 (31)

LOCAL AM & FM STATIONS	FREQUENCY
Radio Denmark News summary.	243 FM 90.8

INTERNATIONAL BROADCASTERS

BBC World Service

198
1296
6180
6195
7120
7325
9410
9660
9750
9760
12095
15070
15575

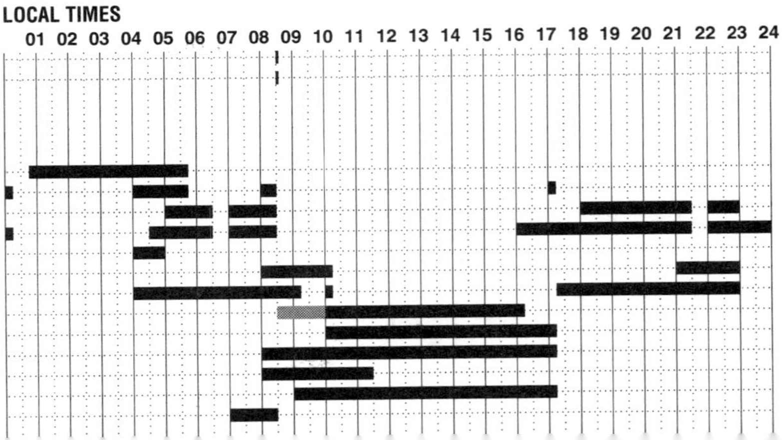

LOCAL TIMES
01 02 03 04 05 06 07 08 09 10 11 12 13 14 15 16 17 18 19 20 21 22 23 24

COPENHAGEN 59

Christian Science Monitor World Service

17640
7510
9495
9840
9985
13770
15665
17510

Radio Canada International

5995
6050
6150
7235
7295
9750
11775
11945
15325
17840
17875

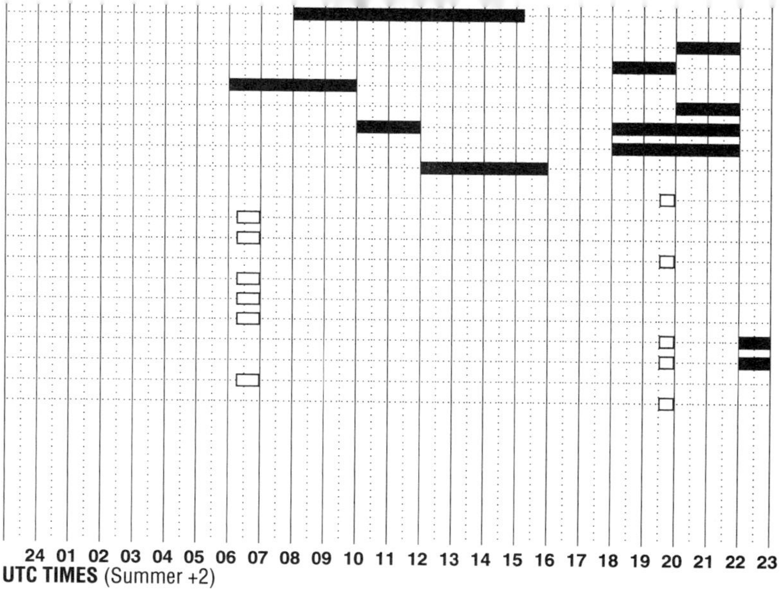

UTC TIMES (Summer +2)

60 COPENHAGEN

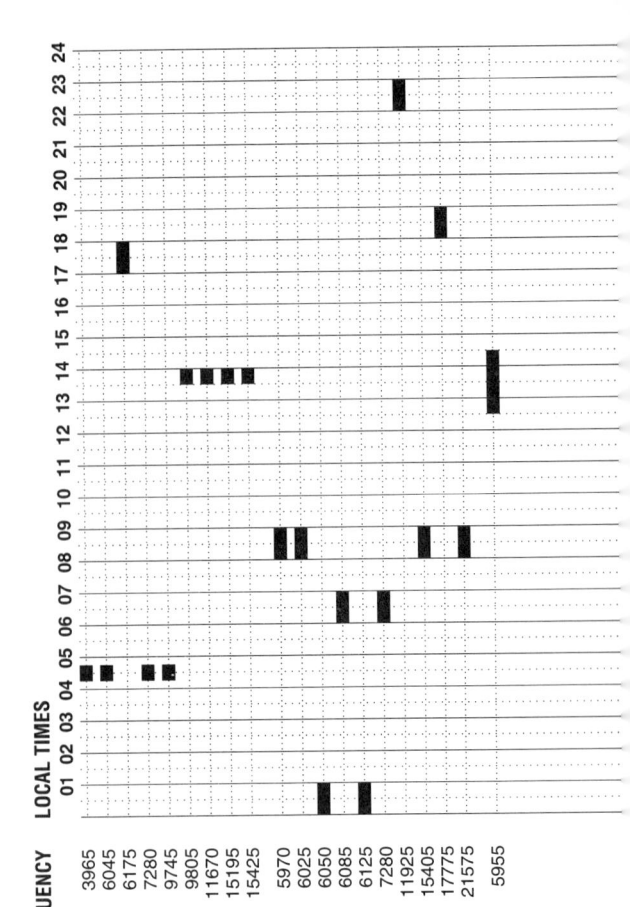

Frequency	Station
3965, 6045, 6175, 7280, 9745, 9805, 11670, 15195, 15425	Radio France International
5970, 6025, 6050, 6085, 6125, 7280, 11925, 15405, 17775, 21575	Radio Japan
5955	Radio Netherlands

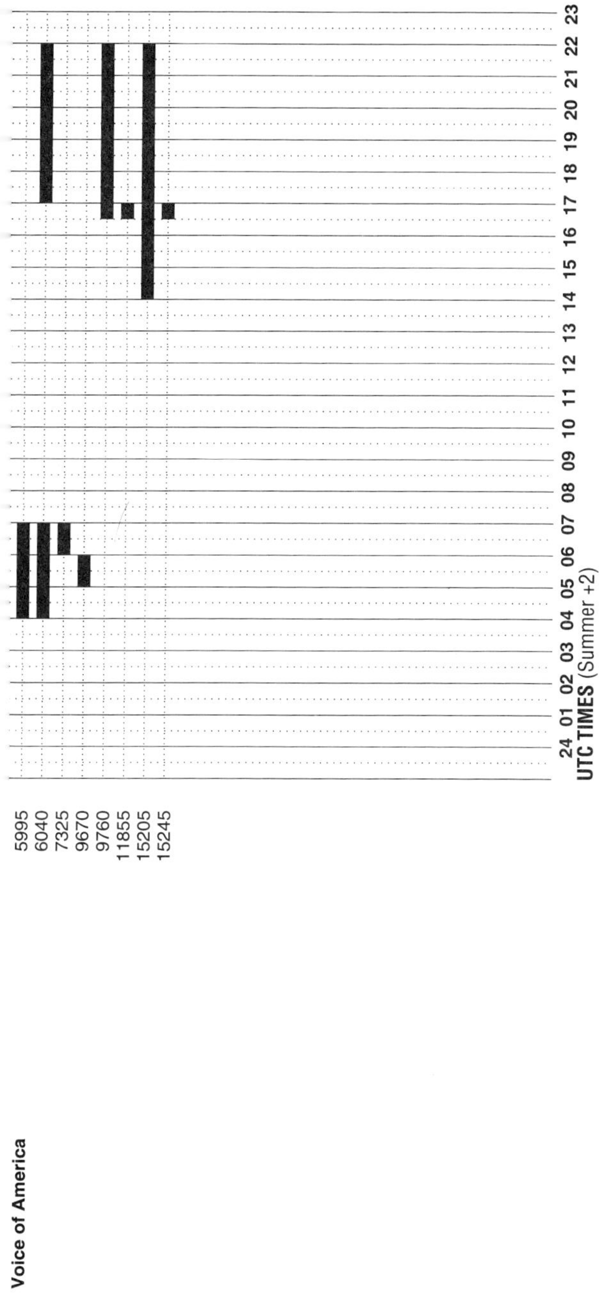

COPENHAGEN 61

Voice of America

5995	
6040	
7325	
9670	
11855	
15205	
15245	

UTC TIMES (Summer +2)

62 DELHI

DELHI, INDIA
Hours difference from UTC (GMT): +5.5
Main business languages: English
Electricity: 50Hz, 220/380V
Currency: Indian Rupee
Telephone country prefix and area code: 91 (11)

LOCAL AM & FM STATIONS	FREQUENCY
All India Radio Program B	666
All India Radio Program A	819
All India Radio Program D	1017
All India Radio Program C	1368
All India Radio FM Program	FM 107.1

INTERNATIONAL BROADCASTERS	
BBC World Service	1413
	5965
	7215
	9580
	9740
	11750
	11955
	15310

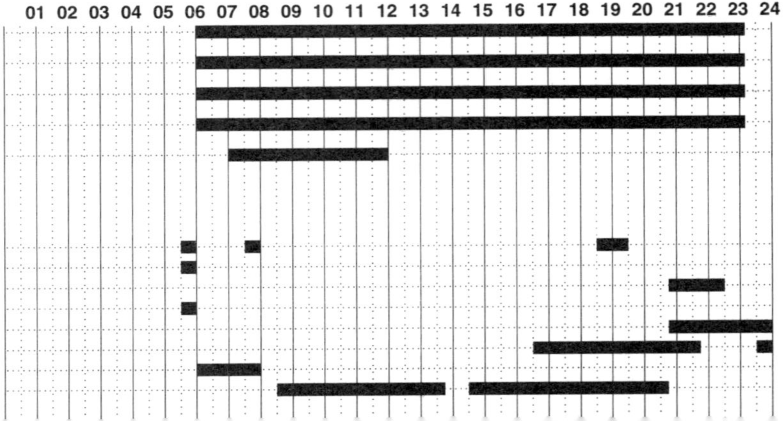

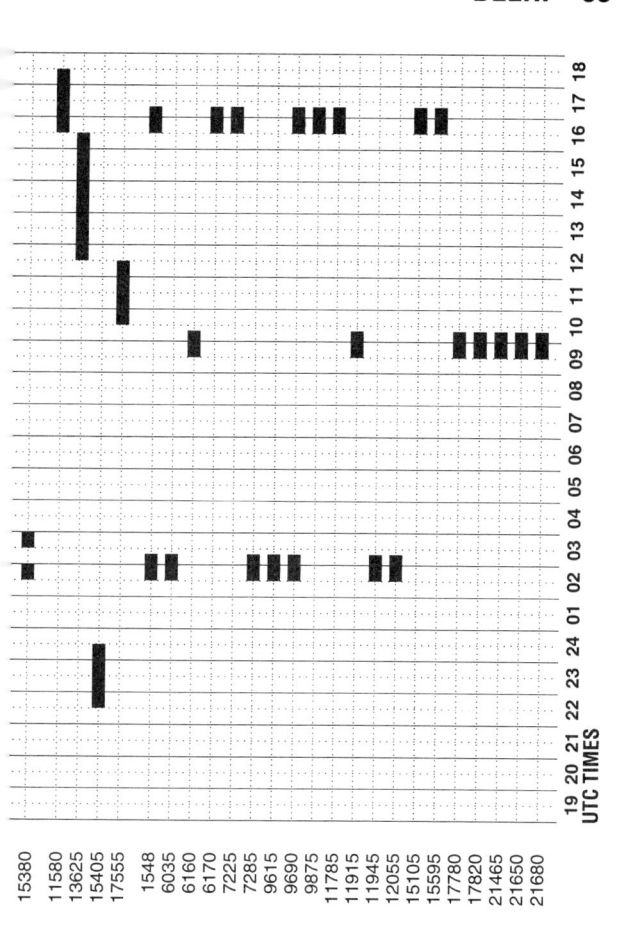

DELHI 63

Christian Science Monitor World Service

15380
11580
13625
15405
17555

Deutsche Welle

1548
6035
6160
6170
7225
7285
9615
9690
9875
11785
11915
11945
12055
15105
15595
17780
17820
21465
21650
21680

UTC TIMES: 19 20 21 22 23 24 01 02 03 04 05 06 07 08 09 10 11 12 13 14 15 16 17 18

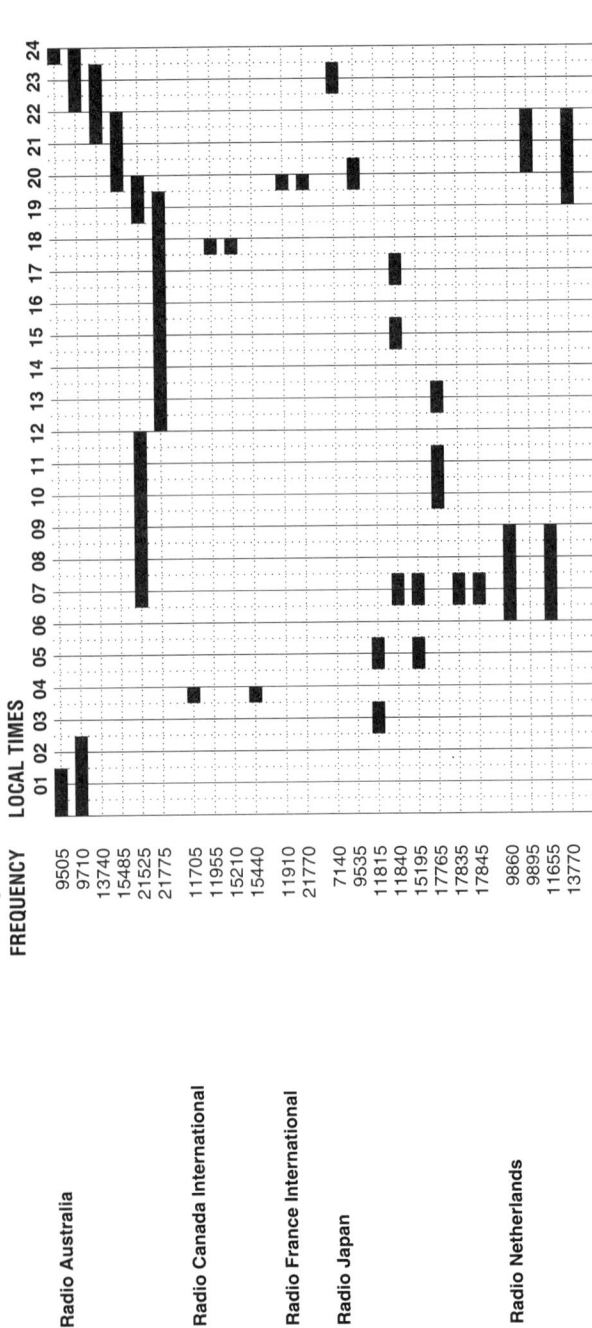

64 DELHI

Station	Frequency
Radio Australia	9505, 9710, 13740, 15485, 21525, 21775
Radio Canada International	11705, 11955, 15210, 15440
Radio France International	11910, 21770
Radio Japan	7140, 9535, 11815, 11840, 15195, 17765, 17835, 17845
Radio Netherlands	9860, 9895, 11655, 13770

DELHI 65

Voice of America

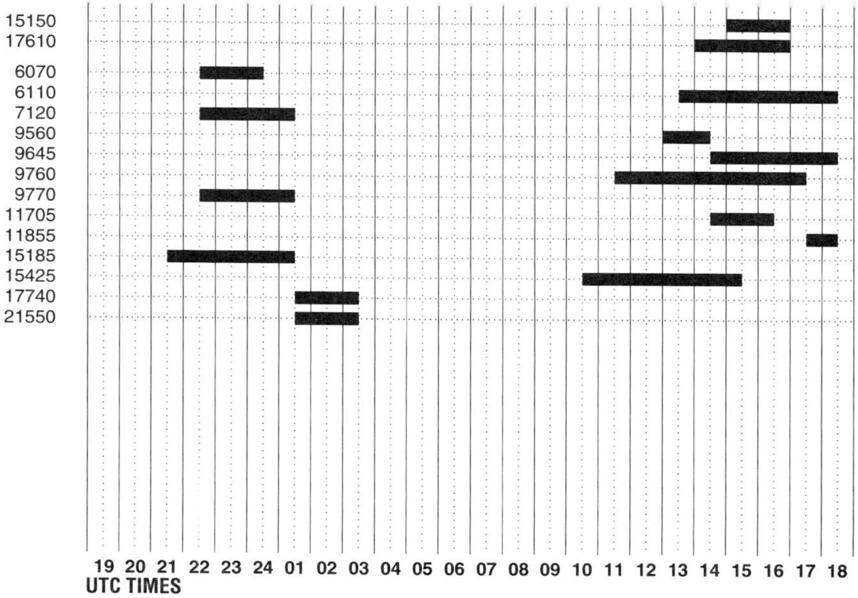

Frequency	UTC Times
15150	15–16
17610	15–16
6070	23–24
6110	13–14, 23–01
7120	00–01
9560	15–16
9645	12–13
9760	15–17
9770	23–01
11705	00–01
11855	11–13
15185	00–01, 15–16
15425	10–16
17740	17–18, 02–03
21550	02–03

UTC TIMES

DUBAI

DUBAI, UNITED ARAB EMIRATES
Hours difference from UTC (GMT): +4
Main business languages: Arabic, English
Electricity: 50Hz, 220V
Currency: UAE Dirham
Telephone country prefix and area code: 971 (4)

LOCAL AM & FM STATIONS

	FREQUENCY
U.A.E. Radio Dubai Government station. News 0730, 0930, 1430, 1730, 2030.	FM 92.0

INTERNATIONAL BROADCASTERS

BBC World Service	639
	702
	720
	1323
	1413
	5965
	6195
	9580
	11760
	11955
Christian Science Monitor World Service	9495

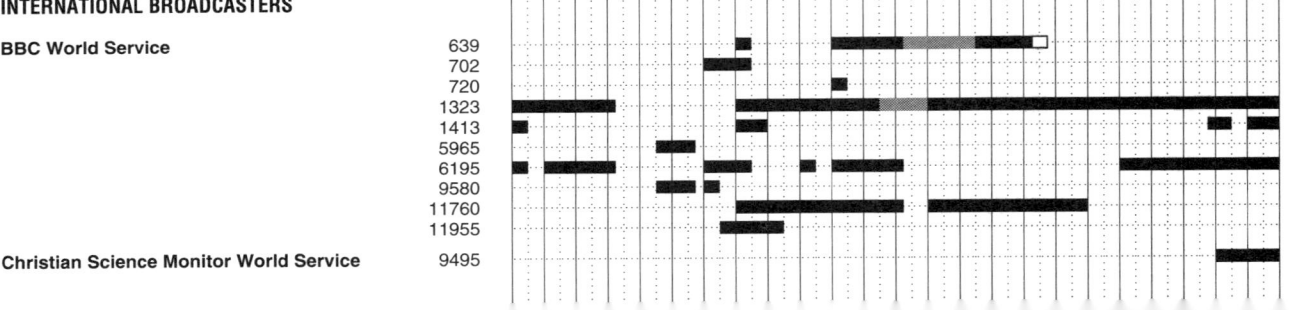

DUBAI 67

Radio Canada International

Radio Japan

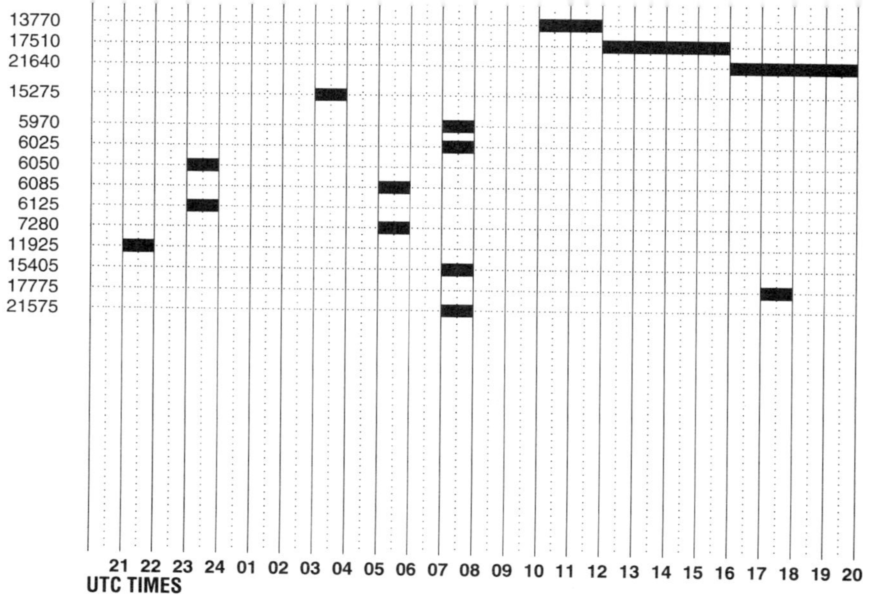

68 DUBAI

Voice of America

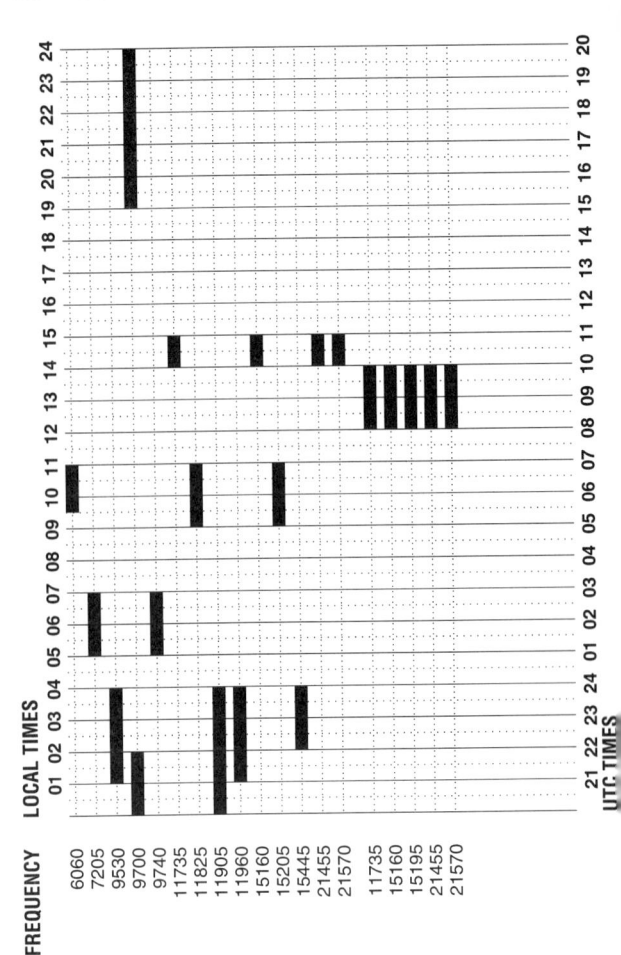

FREQUENCY	Schedule
6060	09–11
7205	05–07
9530	01–04
9700	02–03
9740	06–07
11735	14–15
11825	05–07
11905	01–04
11960	02–04
15160	14–15
15205	05–07
15445	14–15
21455	08–11
21570	19–24

Voice of America - VOA Europe

FREQUENCY	Schedule
11735	
15160	
15195	
21455	
21570	

DUBLIN, IRELAND

Hours difference from UTC (GMT): 0 (Summer +1)
Main business languages: English
Electricity: 50Hz, 220/380V
Currency: Irish Punt
Telephone country prefix and area code: 353 (1)

DUBLIN 69

LOCAL AM & FM STATIONS	FREQUENCY
Atlantic 252 Commercial pop music station.	252
RTE Radio One National network. News on the hour.	567
RTE Radio Two National network. News on the hour.	1278
RTE Radio One National network. News on the hour.	FM 88.5
RTE Radio Two National network. News on the hour.	FM 90.7
RTE FM3	FM 92.9
Classic Hits 98 FM Commercial station.	FM 98.1
FM 104 Commercial station. Contemporary Hits.	FM 104.4

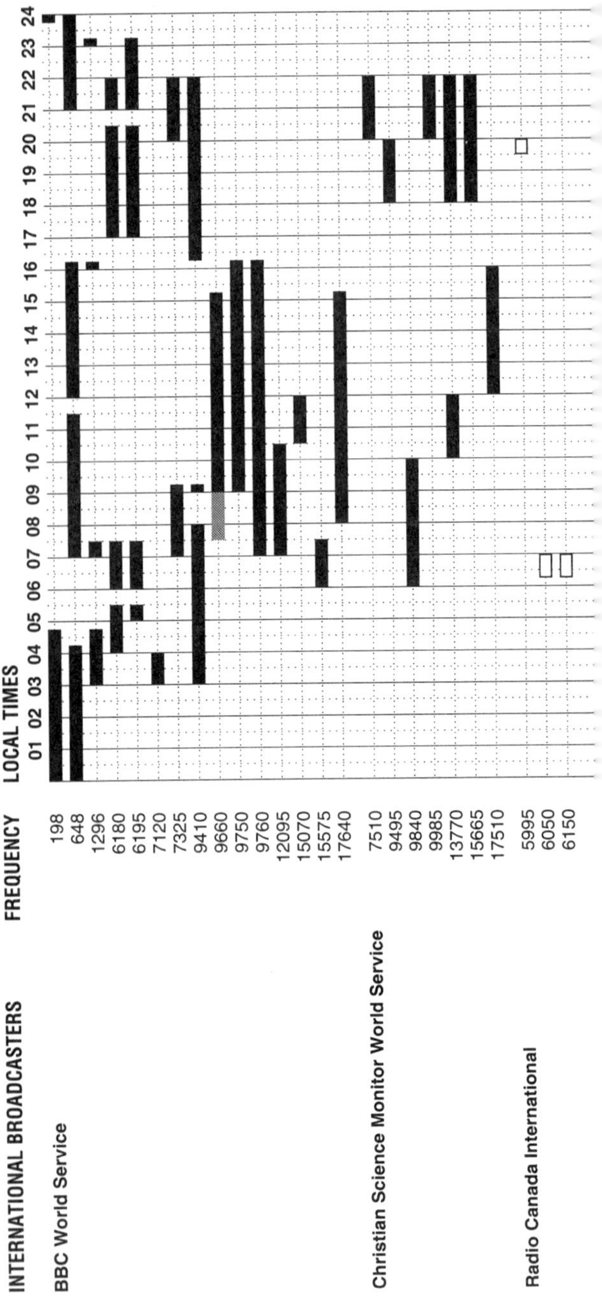

70 DUBLIN

INTERNATIONAL BROADCASTERS	FREQUENCY
BBC World Service	198
	648
	1296
	6180
	6195
	7120
	7325
	9410
	9660
	9750
	9760
	12095
	15070
	15575
	17640
Christian Science Monitor World Service	7510
	9495
	9840
	9985
	13770
	15665
	17510
Radio Canada International	5995
	6050
	6150

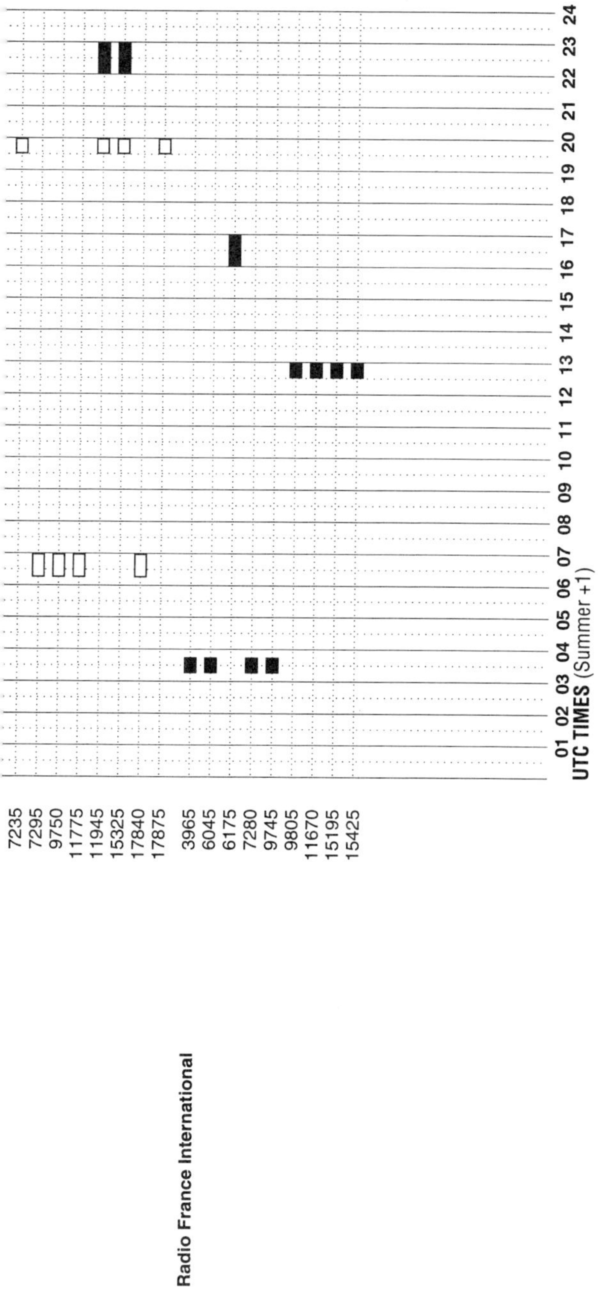

DUBLIN 71

Radio France International

Frequency
7235
7295
9750
11775
11945
15325
17840
17875
3965
6045
6175
7280
9745
9805
11670
15195
15425

UTC TIMES (Summer +1)

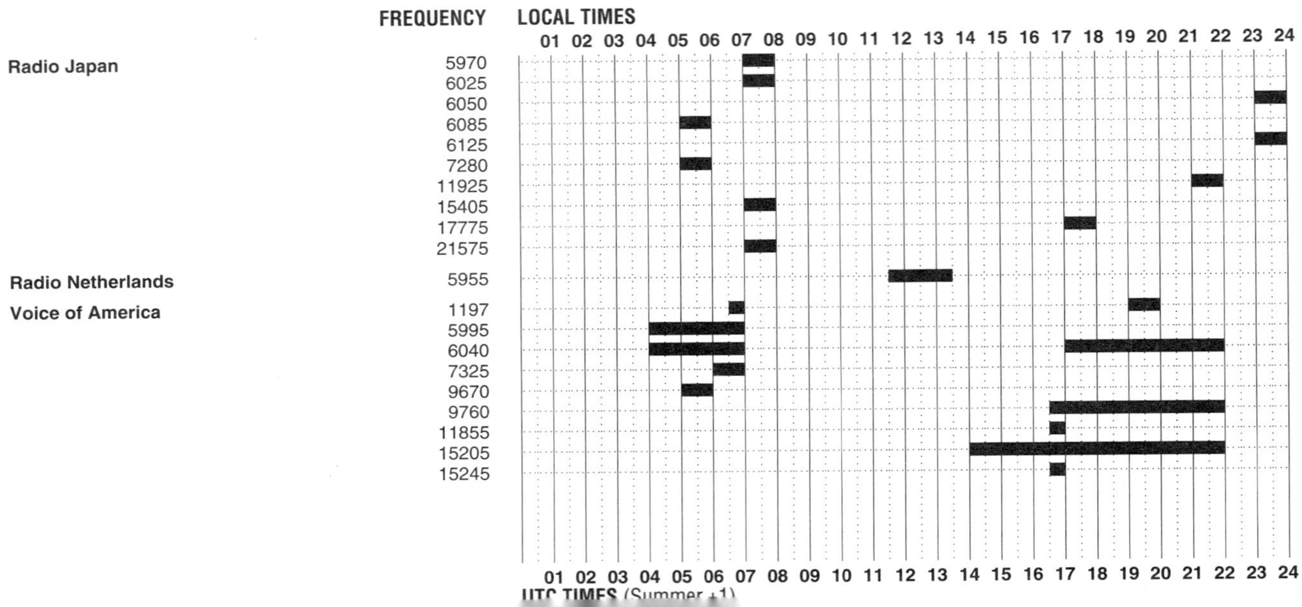

72 DUBLIN

	FREQUENCY	LOCAL TIMES
Radio Japan	5970, 6025, 6050, 6085, 6125, 7280, 11925, 15405, 17775, 21575	
Radio Netherlands	5955	
Voice of America	1197, 5995, 6040, 7325, 9670, 9760, 11855, 15205, 15245	

UTC TIMES (Summer +1)

FRANKFURT, GERMANY

Hours difference from UTC (GMT): +1 (Summer +2)
Main business languages: German, English
Electricity: 50Hz, 220V
Currency: Deutschmark
Telephone country prefix and area code: 49 (69)

LOCAL AM & FM STATIONS	FREQUENCY
American Forces Network News on the hour by satellite from the U.S.	873
Deutschlandfunk Program for listeners abroad.	1269
American Forces Network - AFN2 Easy listening music station.	FM 98.7

FRANKFURT   73

74 FRANKFURT

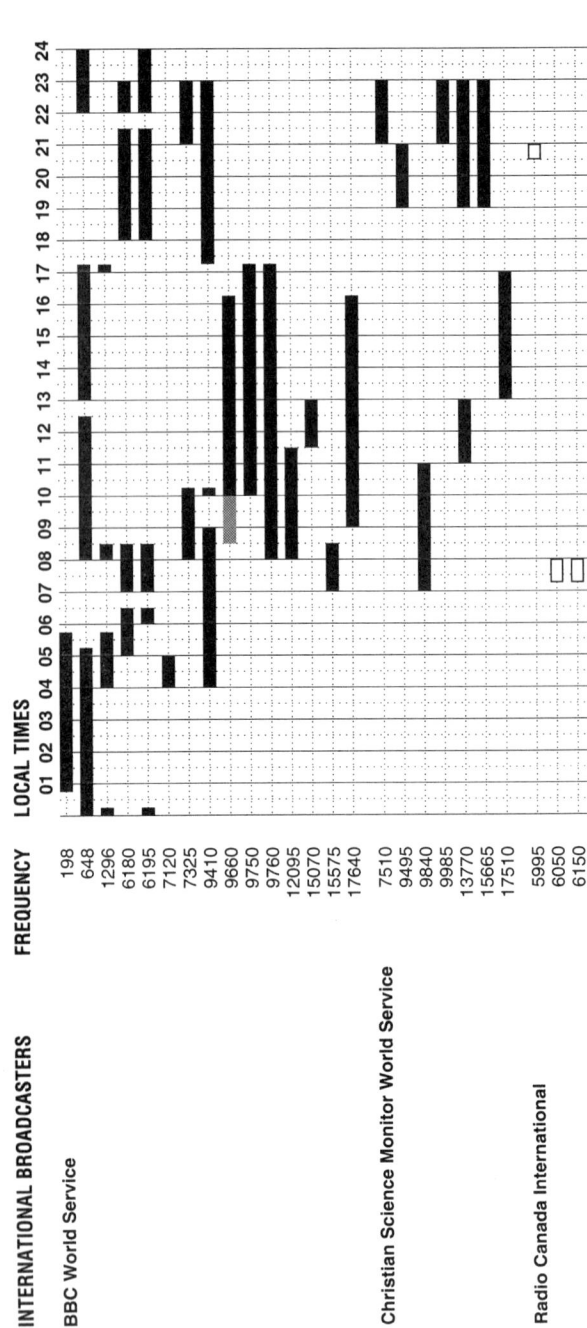

INTERNATIONAL BROADCASTERS	FREQUENCY
BBC World Service	198
	648
	1296
	6180
	6195
	7120
	7325
	9410
	9660
	9750
	9760
	12095
	15070
	15575
	17640
Christian Science Monitor World Service	7510
	9495
	9840
	9985
	13770
	15665
	17510
Radio Canada International	5995
	6050
	6150

FRANKFURT 75

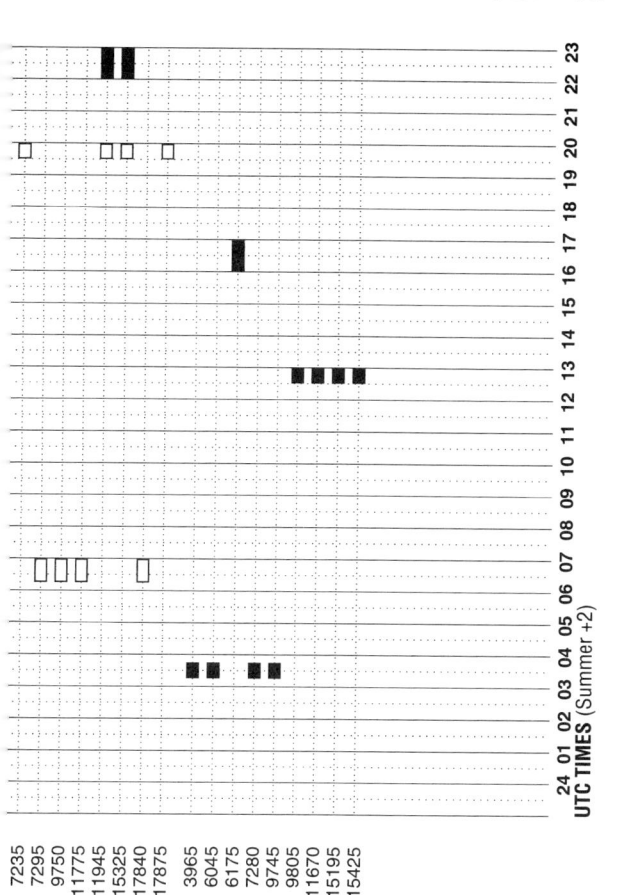

Radio France International

Frequency	Schedule
7235	
7295	
9750	20
11775	
11945	07
15325	07
17840	22
17875	20
3965	03-04
6045	03-04
6175	03
7280	16
9745	03-04
9805	
11670	12-13
15195	12-13
15425	12-13

UTC TIMES (Summer +2)

76 FRANKFURT

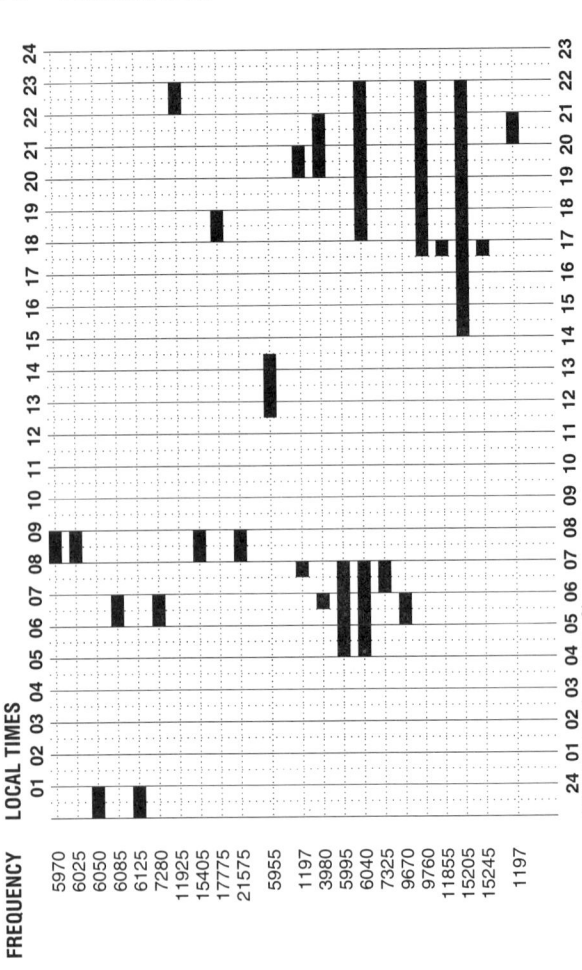

FREQUENCY	
Radio Japan	5970, 6025, 6050, 6085, 6125, 7280, 11925, 15405, 17775, 21575
Radio Netherlands	5955
Voice of America	1197, 3980, 5995, 6040, 7325, 9670, 9760, 11855, 15205, 15245
Voice of America - VOA Europe	1197

LOCAL TIMES / UTC TIMES (Summer +2)

GENEVA, SWITZERLAND

Hours difference from UTC (GMT): +1 (Summer +2)
Main business languages: French, English
Electricity: 50Hz, 220V
Currency: Swiss Franc
Telephone country prefix and area code: 41 (22)

LOCAL AM & FM STATIONS   FREQUENCY

Voice of America - VOA Europe   FM 103.2
Program for young Europeans. News on the
hour. Available on hotel cable system only.

INTERNATIONAL BROADCASTERS

BBC World Service
5965
6195
7120
7325
9410
9760
12095
15070
17640

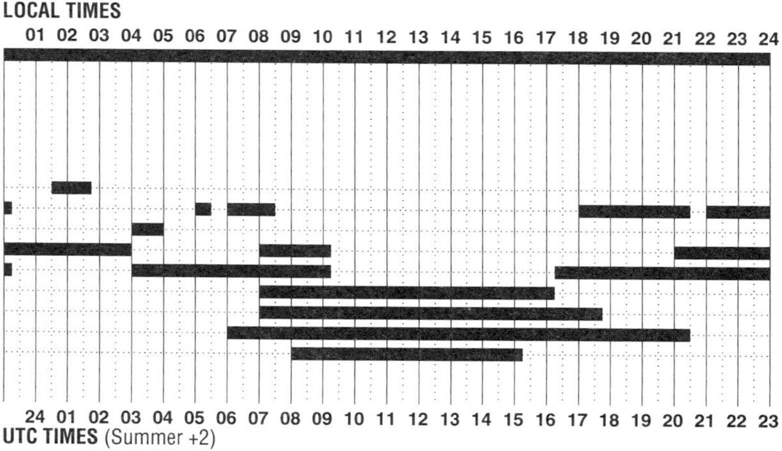

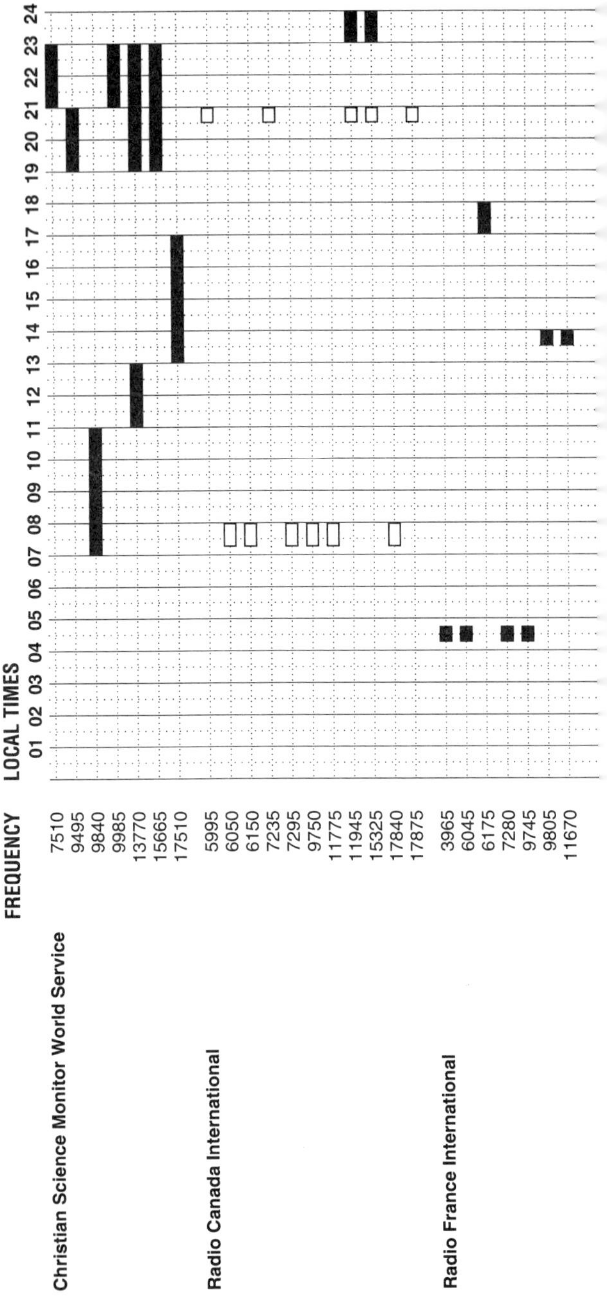

78 GENEVA

GENEVA 79

Radio Japan

15195
15425

5970
6025
6050
6085
6125
7280
11925
15405
17775
21575

Radio Netherlands

5955

Voice of America

5995
6040
7325
9670
9760
11855
15205
15245

UTC TIMES (Summer +2)

80 GLASGOW

GLASGOW, SCOTLAND
Hours difference from UTC (GMT): 0 (Summer +1)
Main business languages: English
Electricity: 50Hz, 240V
Currency: Pound Sterling
Telephone country prefix and area code: 44 (41)

LOCAL AM & FM STATIONS

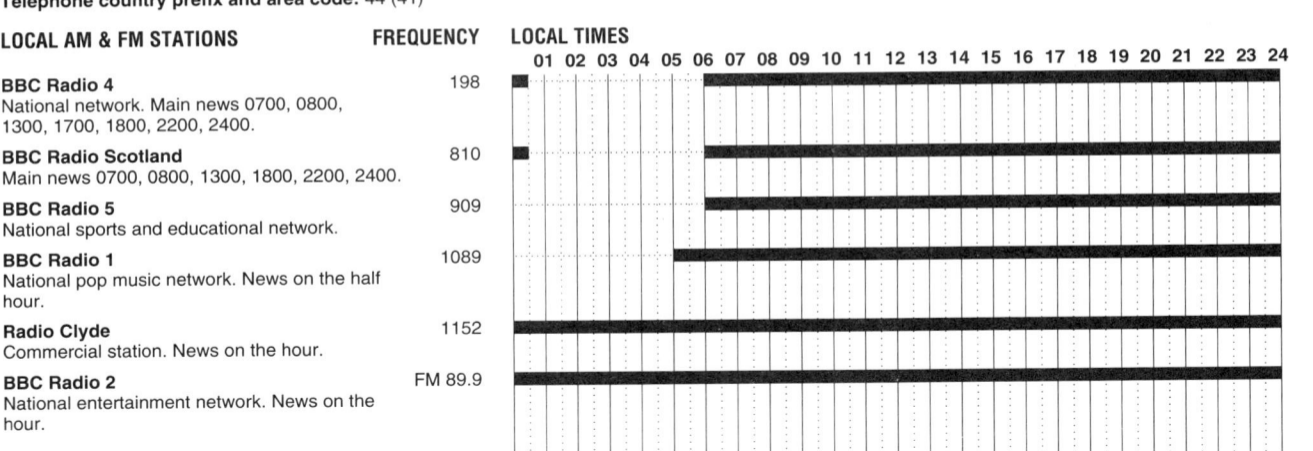

Station	Frequency	Local Times
BBC Radio 4 — National network. Main news 0700, 0800, 1300, 1700, 1800, 2200, 2400.	198	
BBC Radio Scotland — Main news 0700, 0800, 1300, 1800, 2200, 2400.	810	
BBC Radio 5 — National sports and educational network.	909	
BBC Radio 1 — National pop music network. News on the half hour.	1089	
Radio Clyde — Commercial station. News on the hour.	1152	
BBC Radio 2 — National entertainment network. News on the hour.	FM 89.9	

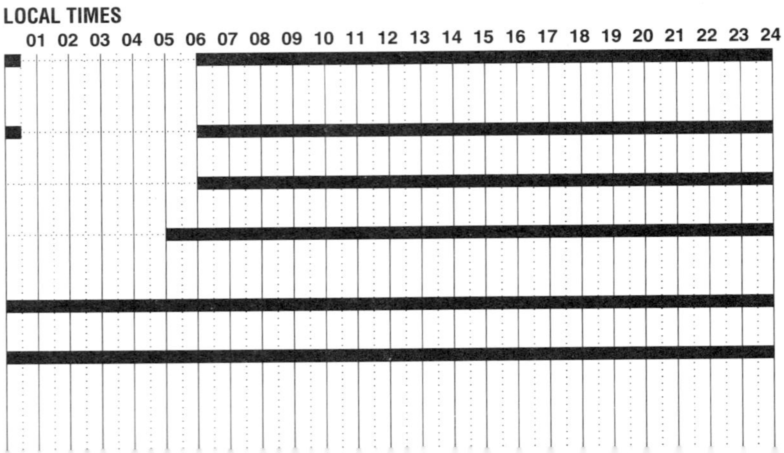

GLASGOW 81

BBC Radio 3
Classical music station.
FM 92.1

BBC Radio 4
National network. Main news 0700, 0800, 1300, 1700, 1800, 2200, 2400.
FM 94.3

BBC Radio Scotland
Main news 0700, 0800, 1300, 1800, 2200, 2400.
FM 95.8

BBC Radio 1
National pop music network. News on the half hour.
FM 99.5

Clyde FM
Commercial station.
FM 102.5

East End Radio
Community station.
FM 103.5

UTC TIMES (Summer +1)

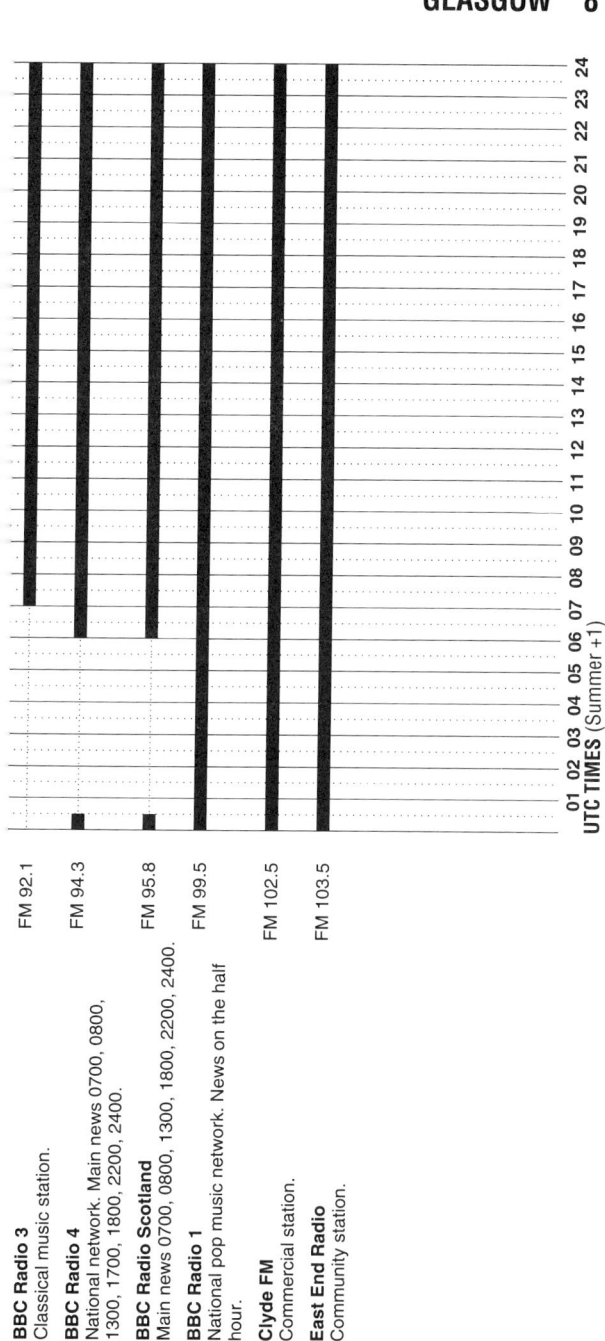

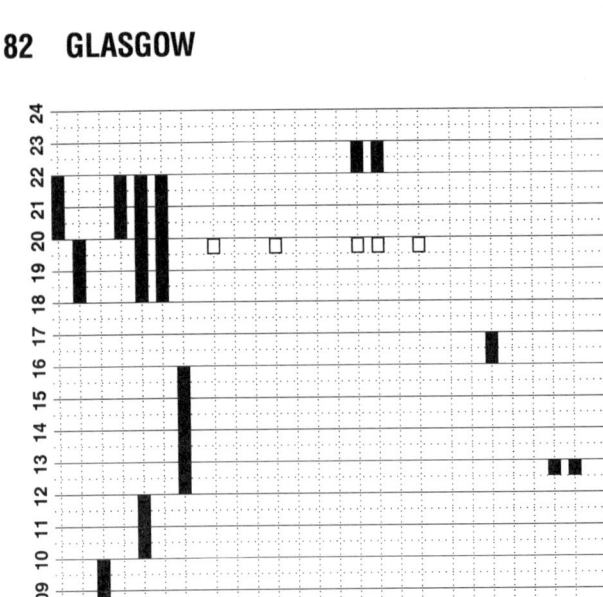

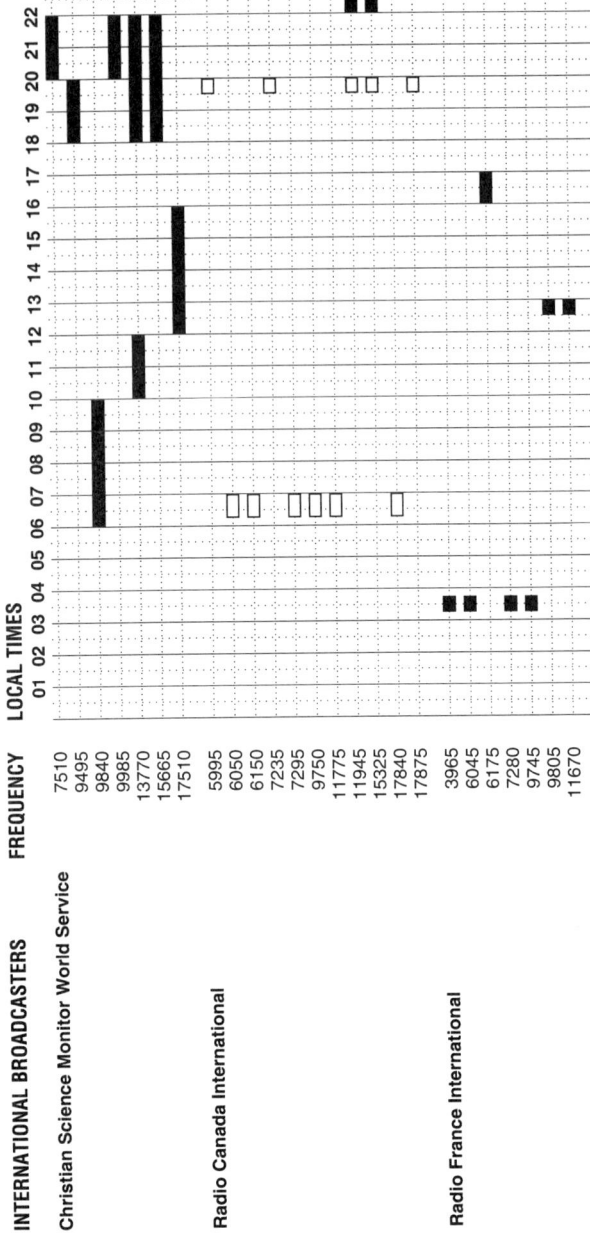

82 GLASGOW

INTERNATIONAL BROADCASTERS	FREQUENCY
Christian Science Monitor World Service	7510
	9495
	9840
	9985
	13770
	15665
	17510
Radio Canada International	5995
	6050
	6150
	7235
	7295
	9750
	11775
	11945
	15325
	17840
	17875
Radio France International	3965
	6045
	6175
	7280
	9745
	9805
	11670

GLASGOW 83

Radio Japan

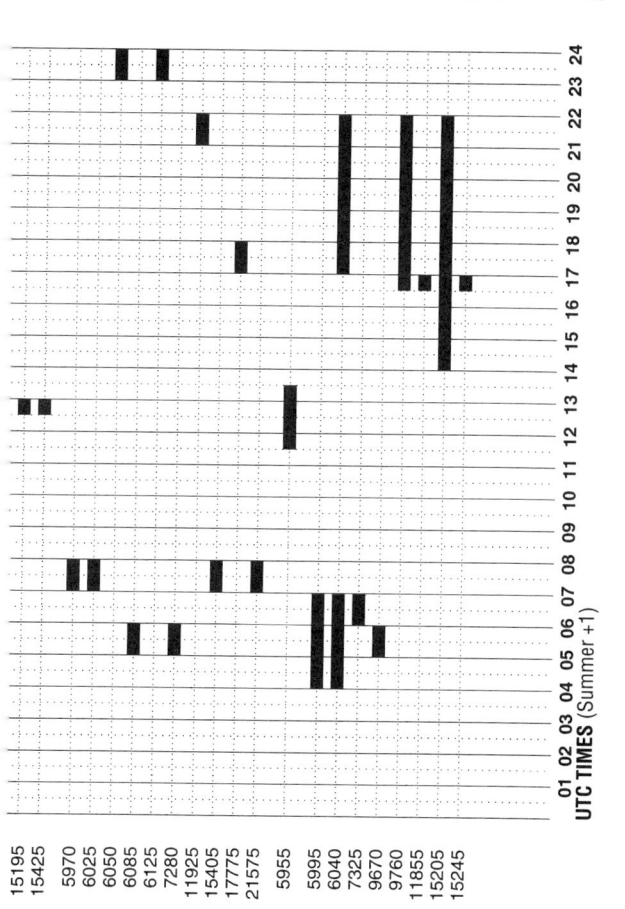

Radio Netherlands

Voice of America

84 HAMBURG

HAMBURG, GERMANY
Hours difference from UTC (GMT): +1 (Summer +2)
Main business languages: German, English
Electricity: 50Hz, 220V
Currency: Deutschmark
Telephone country prefix and area code: 49 (40)

LOCAL AM & FM STATIONS	FREQUENCY
Deutschlandfunk Program for listeners abroad.	1269

INTERNATIONAL BROADCASTERS

BBC World Service

- 198
- 648
- 1296
- 6180
- 6195
- 7120
- 7325
- 9410
- 9660
- 9750
- 9760
- 12095
- 15070

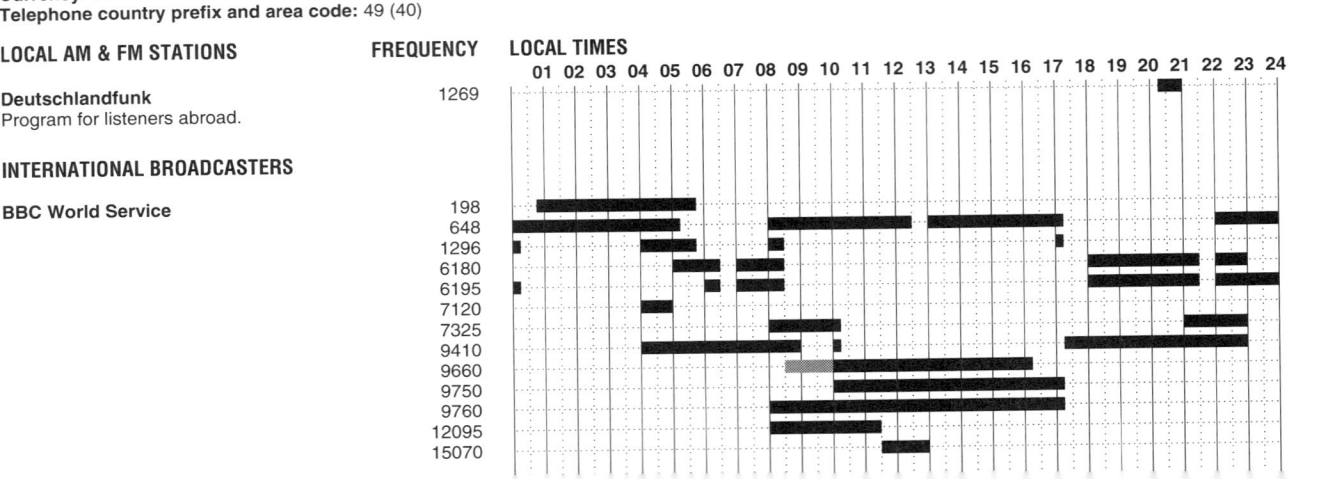

HAMBURG 85

Christian Science Monitor World Service

15575
17640
7510
9495
9840
9985
13770
15665
17510

Radio Canada International

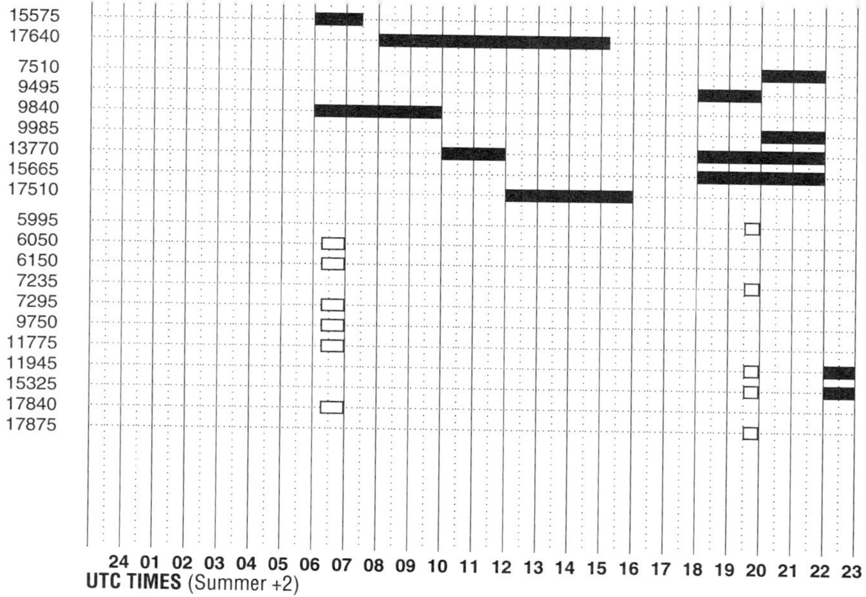

5995
6050
6150
7235
7295
9750
11775
11945
15325
17840
17875

24 01 02 03 04 05 06 07 08 09 10 11 12 13 14 15 16 17 18 19 20 21 22 23
UTC TIMES (Summer +2)

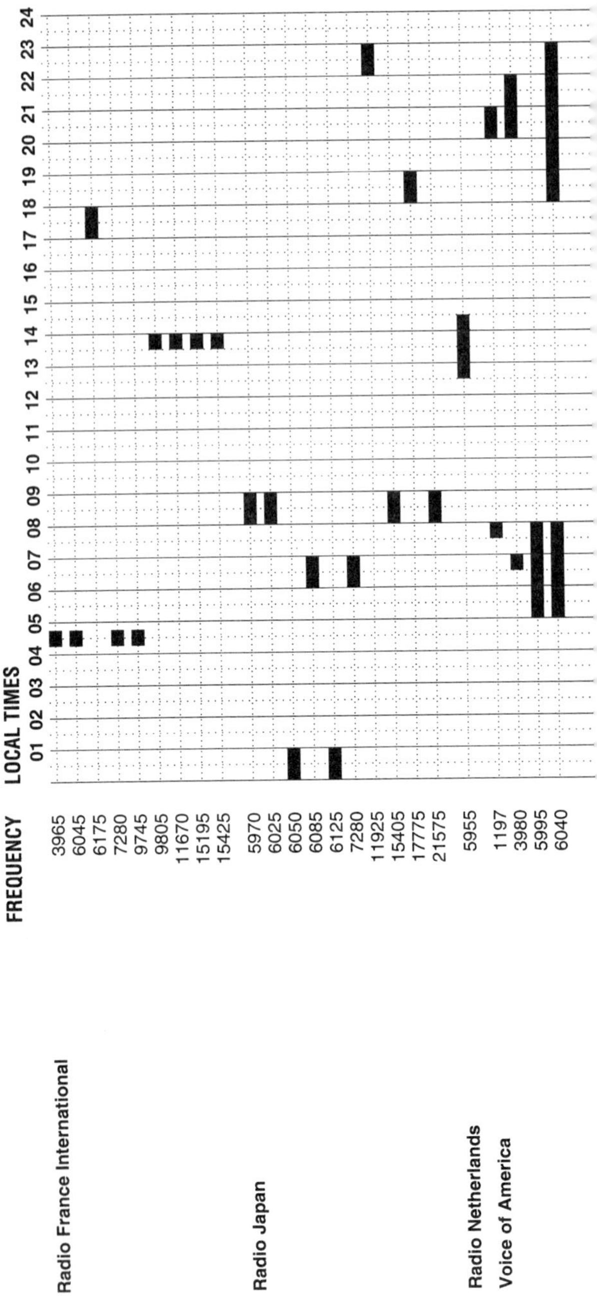

86 HAMBURG

Station	Frequency (kHz)	Local Times
Radio France International	3965	04–05
	6045	04–05
	6175	16–17
	7280	04–05
	9745	04–05
	9805	
	11670	
	15195	
	15425	
Radio Japan	5970	08–09
	6025	08–09
	6050	02–03
	6085	06–07
	6125	02–03
	7280	06–07
	11925	13–14
	15405	08–09
	17775	18–19
	21575	08–09
Radio Netherlands	5955	12–14
Voice of America	1197	20–21
	3980	07–08, 21–22
	5995	06–07
	6040	06–07, 22–23

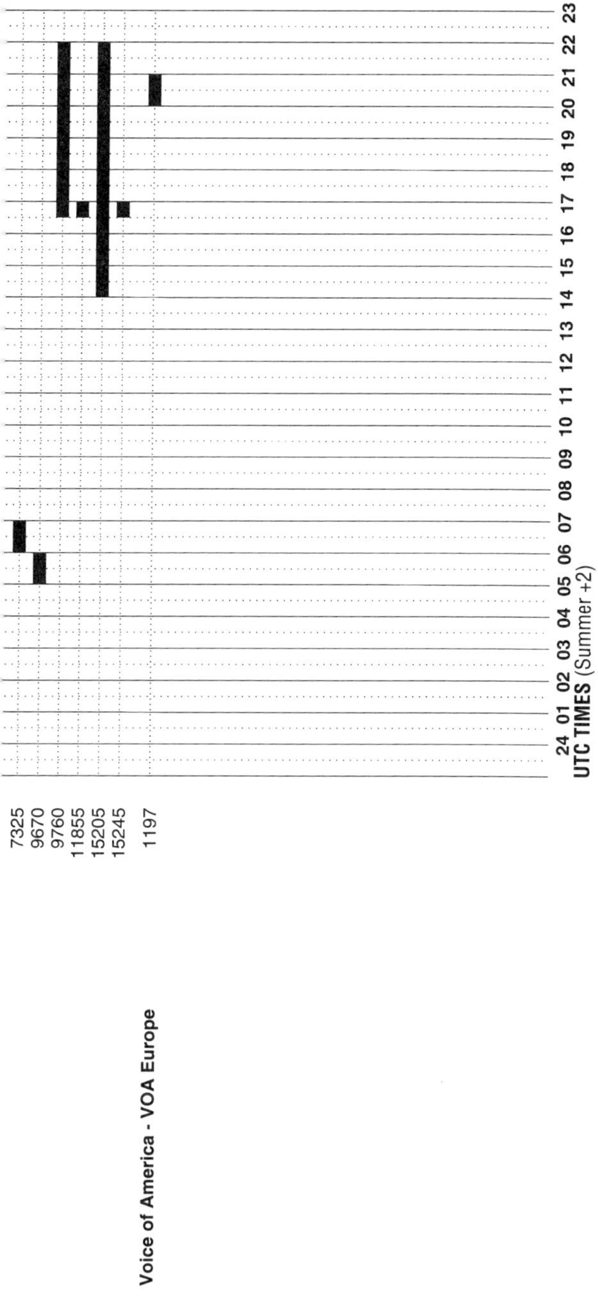

HAMBURG 87

7325	
9670	
11855	
15205	
15245	
1197	

UTC TIMES (Summer +2)

Voice of America - VOA Europe

88 HELSINKI

HELSINKI, FINLAND
Hours difference from UTC (GMT): +2 (Summer +3)
Main business languages: Finnish, English
Electricity: 50Hz, 220V
Currency: Finnish Mark
Telephone country prefix and area code: 358 (0)

LOCAL AM & FM STATIONS | FREQUENCY

Radio Finland — 558
Nordic news at start of broadcast.

Radio Finland - YLE3 — FM 94.7
World news and local weather forecast.

Radio Finland Capital FM — FM 103.7
Nordic news at start of broadcast.

INTERNATIONAL BROADCASTERS

BBC World Service
1296
6180
6195
7120
7325
9410
9660
9750

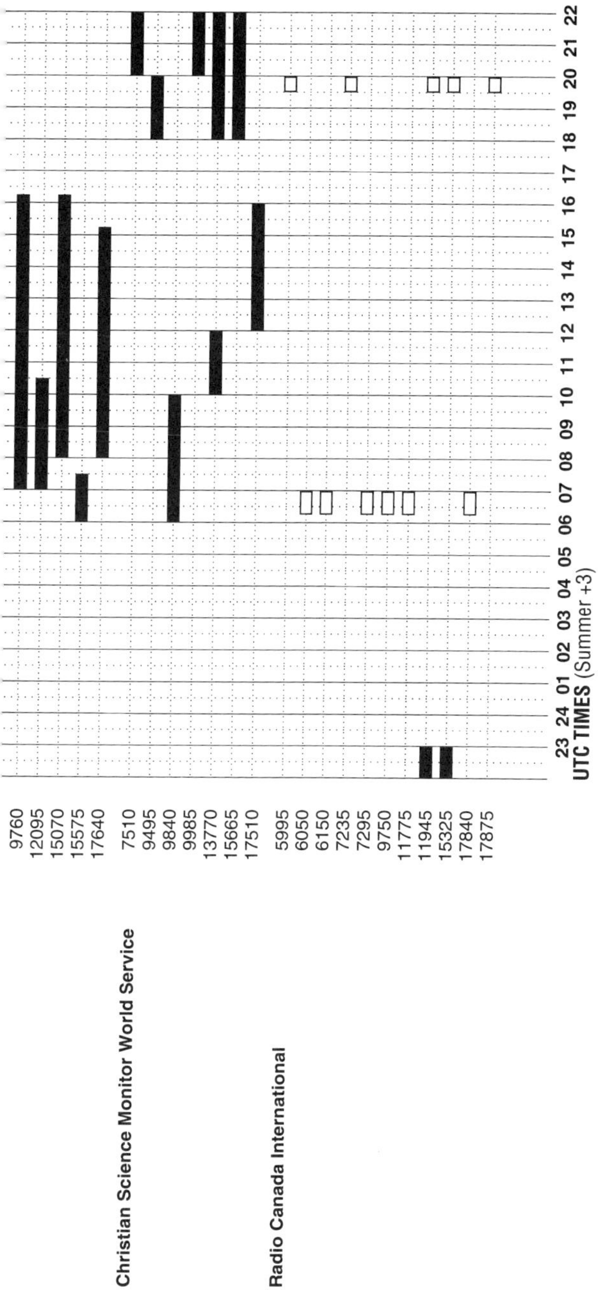

HELSINKI 89

Christian Science Monitor World Service

9760
12095
15070
15575
17640
7510
9495
9840
9985
13770
15665
17510

Radio Canada International

5995
6050
6150
7235
7295
9750
11775
11945
15325
17840
17875

UTC TIMES (Summer +3)

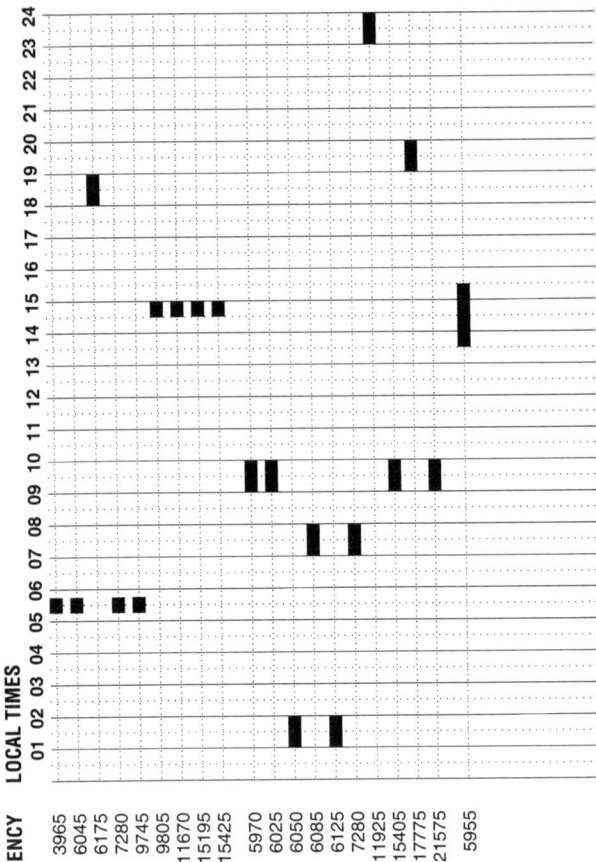

90 HELSINKI

Station	Frequency	Local Times
Radio France International	3965	05–06
	6045	05–06
	6175	18–19
	7280	05–06
	9745	05–06
	9805	
	11670	14–15
	15195	14–15
	15425	14–15
Radio Japan	5970	
	6025	
	6050	01–02
	6085	
	6125	02–03
	7280	
	11925	06–07
	15405	07–08
	17775	09–10, 19–20
	21575	09–10
Radio Netherlands	5955	13–15, 23–24

HELSINKI 91

Voice of America

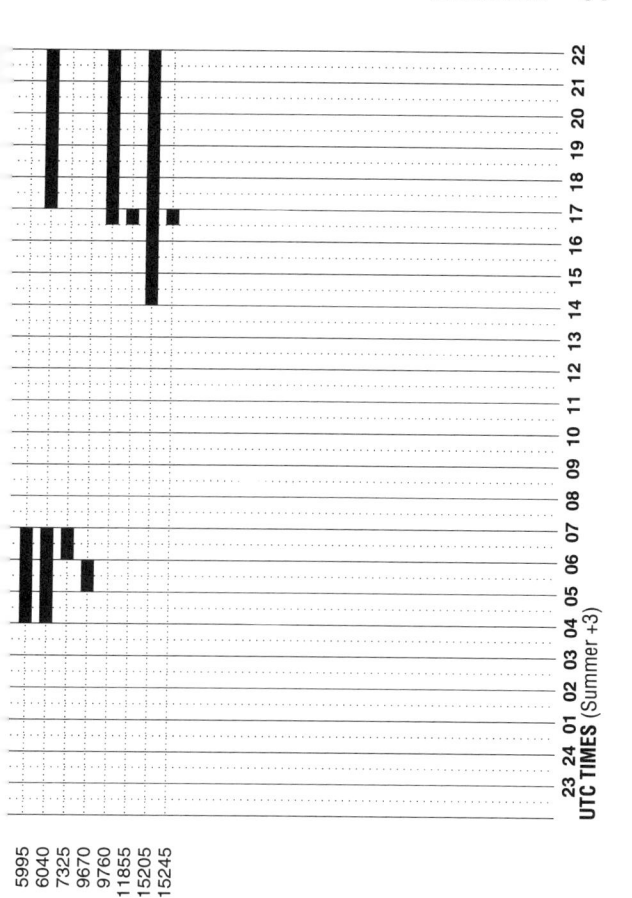

kHz
5995
6040
7325
9670
11855
15205
15245

UTC TIMES (Summer +3)

HONG KONG

Hours difference from UTC (GMT): +8
Main business languages: Chinese, English
Electricity: 50Hz, 200V
Currency: Hong Kong Dollar
Telephone country prefix and area code: 852 (3)

LOCAL AM & FM STATIONS	FREQUENCY
RTHK Radio 3 News on the hour. Relay BBC 0600, 0630, 0700, 0800, 1900, 2100.	567
RTHK Radio 6 Relay of BBC World Service.	675
RTHK Radio 5 Bilingual service in English & Chinese	783
HK Commercial Radio News every half hour.	864
Metro News All news commercial station.	1044
RTHK Radio 4 Bilingual service in English & Chinese.	FM 97.6
Metro FM Select Easy listening commercial station.	FM 104.0

HONG KONG 93

INTERNATIONAL BROADCASTERS

BBC World Service

7180
9570
9740
11765
11820
11945
11955
15280
15360
17790
17830
21715

Christian Science Monitor World Service

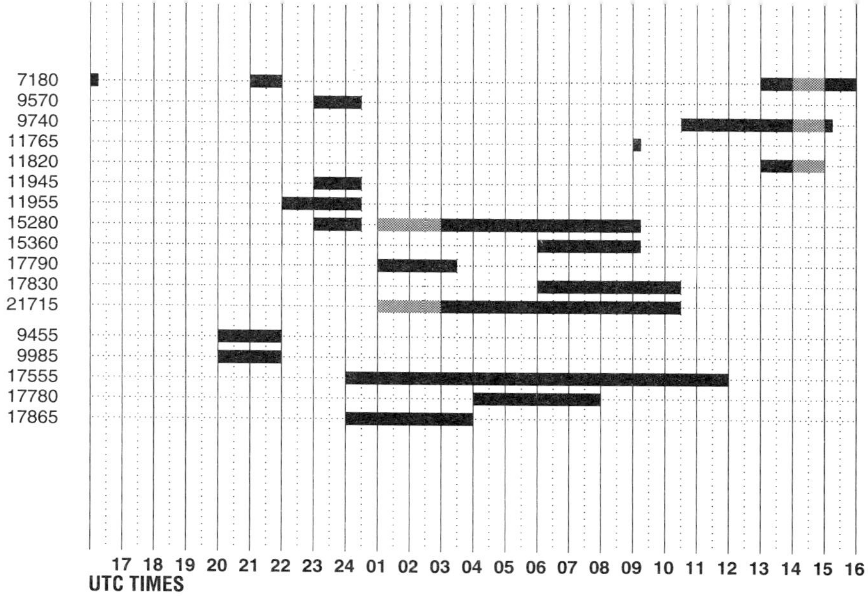

9455
9985
17555
17780
17865

UTC TIMES

94 HONG KONG

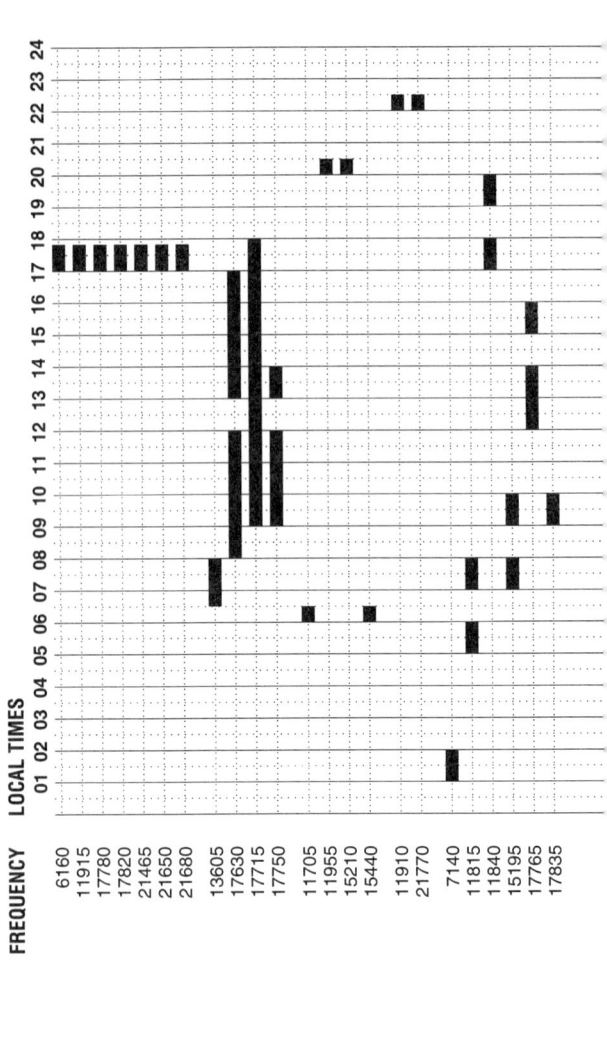

Station	Frequency	Local Times
Deutsche Welle	6160	17–18
	11915	17–18
	17780	17–18
	17820	17–18
	21465	17–18
	21650	17–18
	21680	17–18
Radio Australia	13605	06–08
	17630	13–15
	17715	08–12, 16–18
	17750	08–11, 13–14
Radio Canada International	11705	06–07
	11955	20–21
	15210	20–21
	15440	06–07
Radio France International	11910	22–23
	21770	22–23
Radio Japan	7140	01–02
	11815	05–06
	11840	19–20
	15195	07–08
	17765	07–08
	17835	09–10

HONG KONG 95

Radio Netherlands
- 9630
- 9720
- 11895

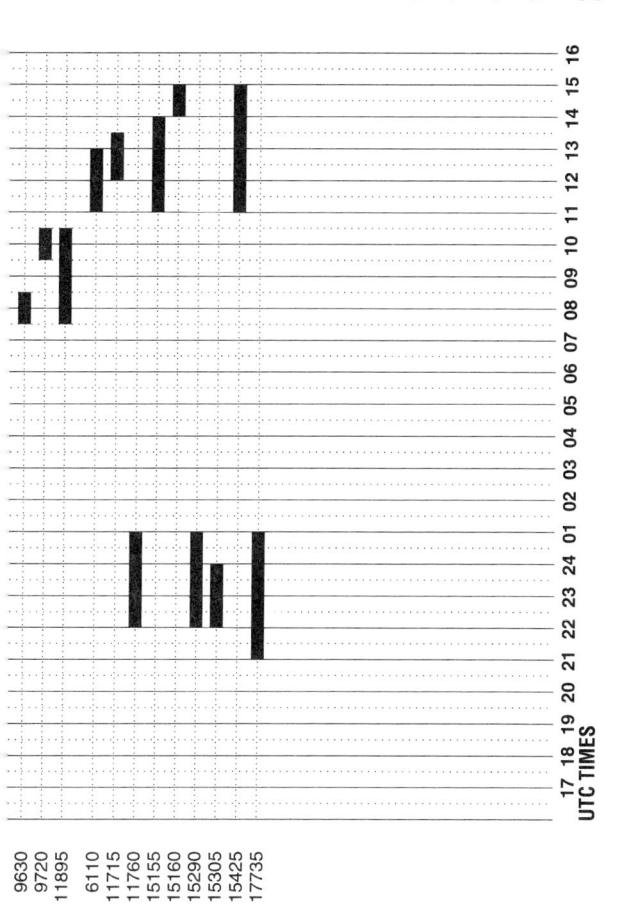

Voice of America
- 6110
- 11715
- 11760
- 15155
- 15160
- 15290
- 15305
- 15425
- 17735

UTC TIMES

96 KUALA LUMPUR

KUALA LUMPUR, MALAYSIA
Hours difference from UTC (GMT): +8
Main business languages: Malay, English
Electricity: 50Hz, 240V
Currency: Ringgit
Telephone country prefix and area code: 60 (3)

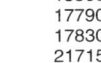

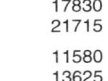

INTERNATIONAL BROADCASTERS	FREQUENCY
BBC World Service	3915
	5975
	6195
	9570
	9740
	11750
	11955
	15310
	15340
	15360
	17790
	17830
	21715
Christian Science Monitor World Service	11580
	13625
	15405
	15610
	17555

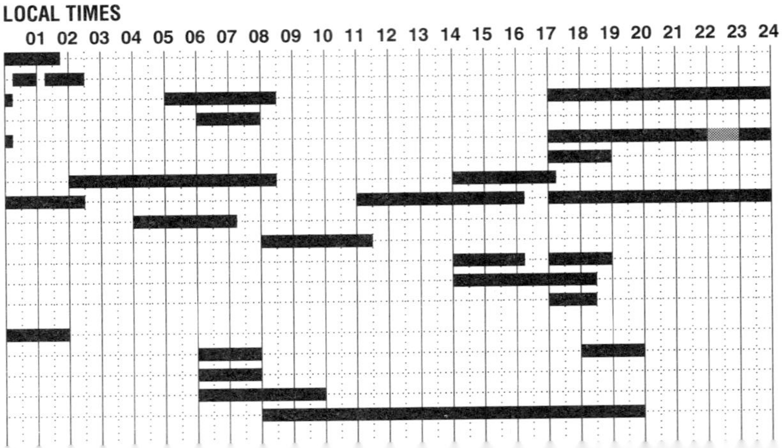

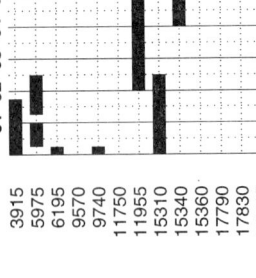

KUALA LUMPUR 97

Deutsche Welle

98 KUALA LUMPUR

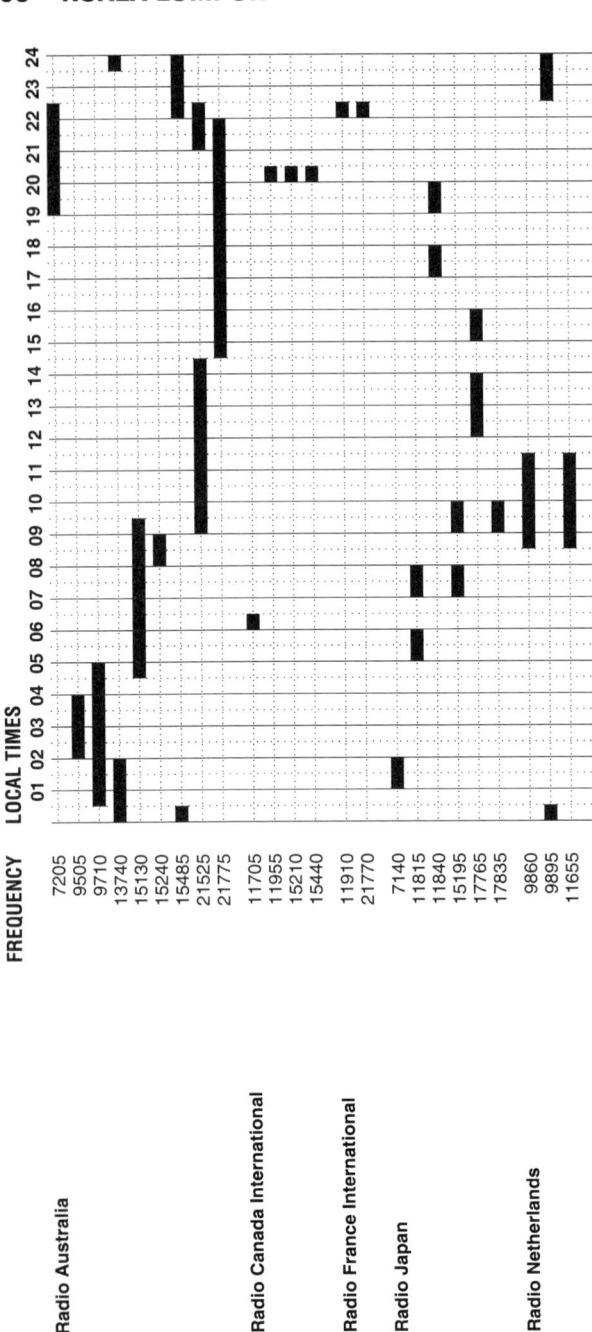

Station	Frequency	Local Times
Radio Australia	7205	19–22
	9505	02–04
	9710	04–05
	13740	01–02
	15130	04–09
	15240	08–09
	15485	01
	21525	12–14, 20–22
	21775	14–16, 23
Radio Canada International	11705	06
	11955	19–20
	15210	20
	15440	20
Radio France International	11910	22
	21770	22
Radio Japan	7140	05
	11815	07
	11840	07–08
	15195	19
	17765	15–16
	17835	18–19
Radio Netherlands	9860	09–10
	9895	01, 09–10
	11655	10–11, 22–23

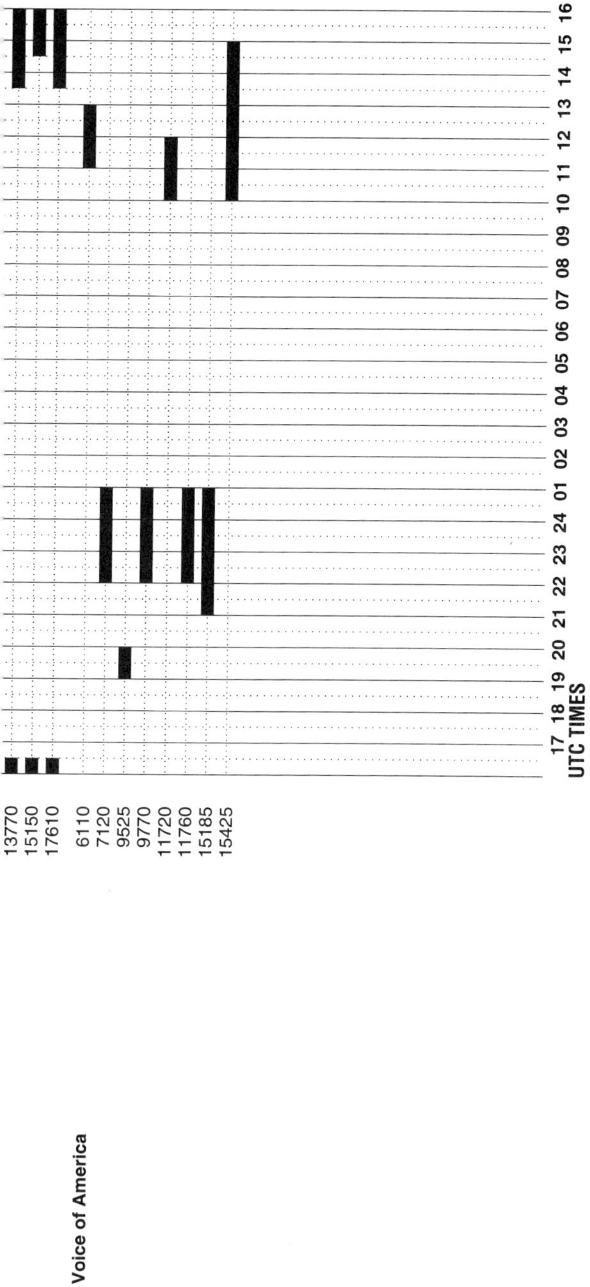

KUALA LUMPUR 99

Voice of America

Frequency (kHz)	UTC Times
13770	17–19
15150	17–19
17610	17–19
6110	11–12
7120	19–20
9525	22–24
9770	11–12
11720	22–24
11760	22–23
15185	21–24
15425	10–15

100 LAGOS

LAGOS, NIGERIA
Hours difference from UTC (GMT): +1
Main business languages: English, Hausa
Electricity: 50Hz, 230/400V
Currency: Naira
Telephone country prefix and area code: 234 (1)

LOCAL AM & FM STATIONS | FREQUENCY

Radio Nigeria Channel One — 1089
News on the hour.

Radio Nigeria Channel Two — 1458
News on the half hour.

Radio Nigeria Channel Three — FM 92.9
Programs in English and Nigerian languages.

INTERNATIONAL BROADCASTERS

BBC World Service
7105
9600
9610
11860
15105
15400
17790
17860

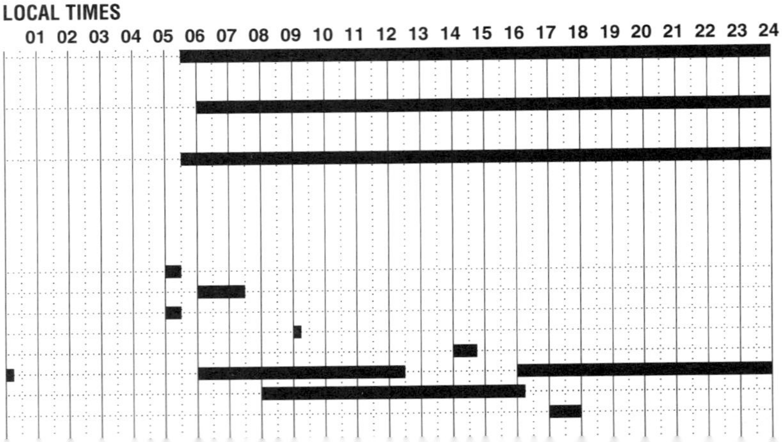

LOCAL TIMES: 01 02 03 04 05 06 07 08 09 10 11 12 13 14 15 16 17 18 19 20 21 22 23 24

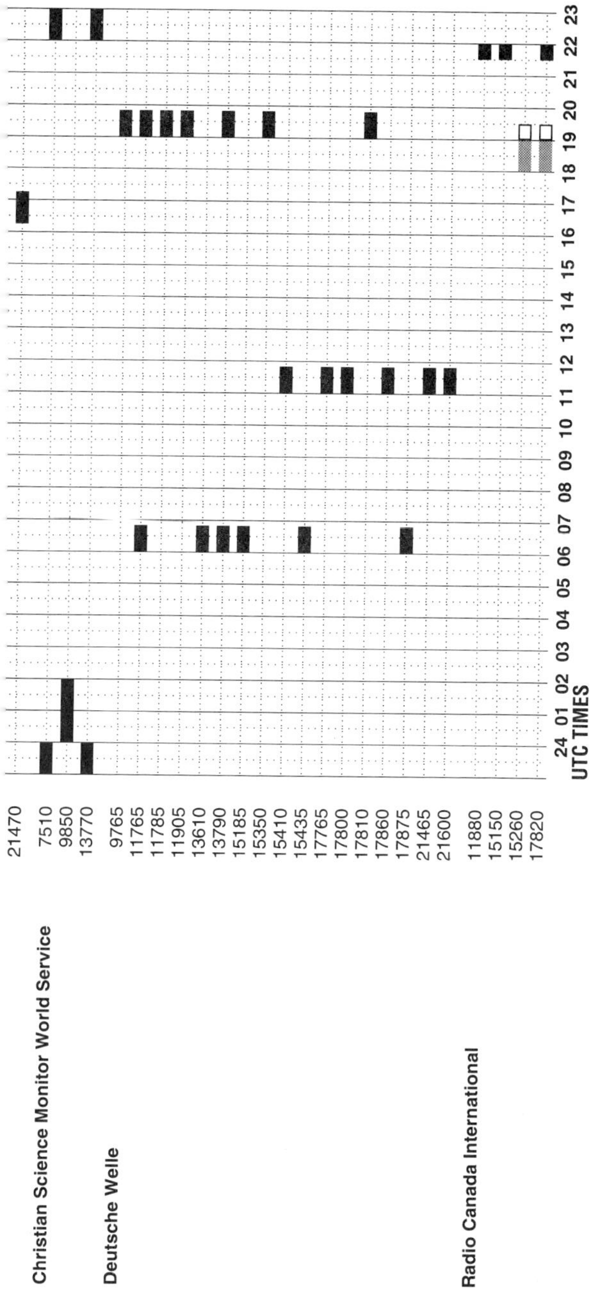

LAGOS 101

Station	Frequencies (kHz)
Christian Science Monitor World Service	21470, 7510, 9850, 13770
Deutsche Welle	9765, 11765, 11785, 11905, 13610, 13790, 15185, 15350, 15410, 15435, 17765, 17800, 17810, 17860, 17875, 21465, 21600
Radio Canada International	11880, 15150, 15260, 17820

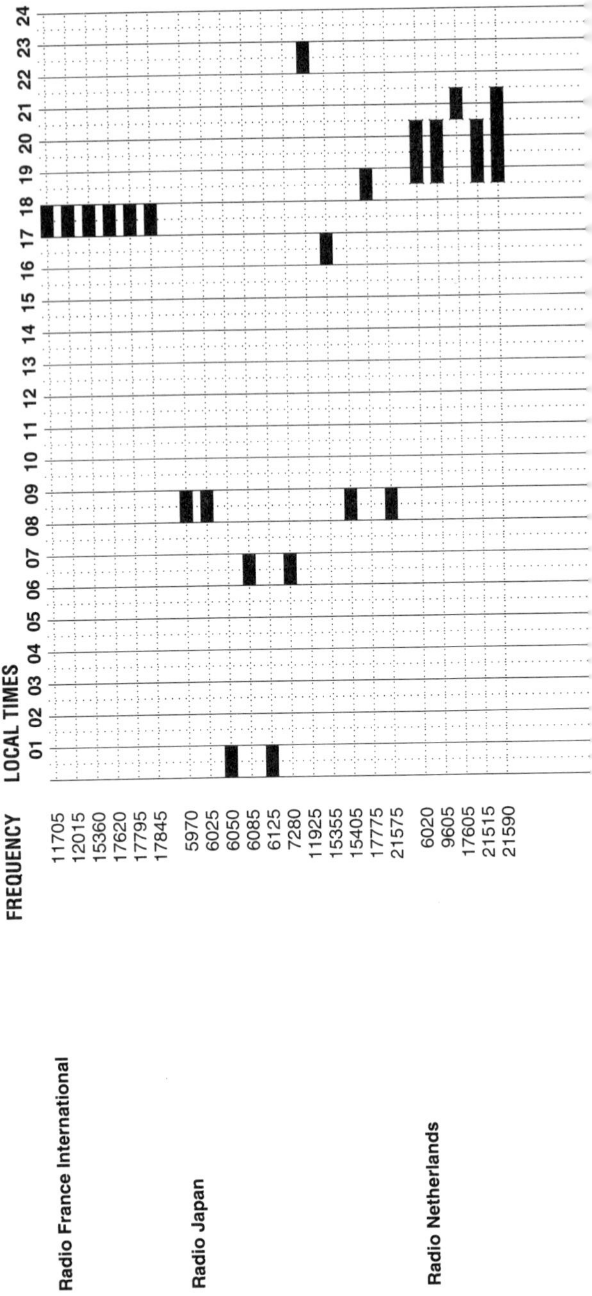

102 LAGOS

FREQUENCY		
Radio France International	11705, 12015, 15360, 17620, 17795, 17845	17–18
Radio Japan	5970, 6025	
	6050	01–02
	6085, 6125	06–07
	7280	
	11925	
	15355, 15405	08–09
	17775, 21575	08–09
		16, 18–19
Radio Netherlands	6020, 9605	19–21
	17605	20–21
	21515	19–21
	21590	18–22

LAGOS 103

Voice of America

104 LONDON

LONDON, UNITED KINGDOM
Hours difference from UTC (GMT): 0 (Summer +1)
Main business languages: English
Electricity: 50Hz, 240V
Currency: Pound Sterling
Telephone country prefix and area code: 44 (71)

LOCAL AM & FM STATIONS	FREQUENCY	LOCAL TIMES 01–24
Spectrum Radio — Programs for ethnic minorities in English and foreign languages.	558	
BBC Radio 4 — Main news 0700, 0800, 1300, 1700, 1800, 2200, 2400.	720	
BBC Radio 5 — Sports and education network.	909	
BBC Radio 1 — Pop music station. News on the half hour.	1089	
London Talkback — Commercial talk station. News on the hour.	1152	
BBC GLR — Music and information service for Greater London area.	1458	

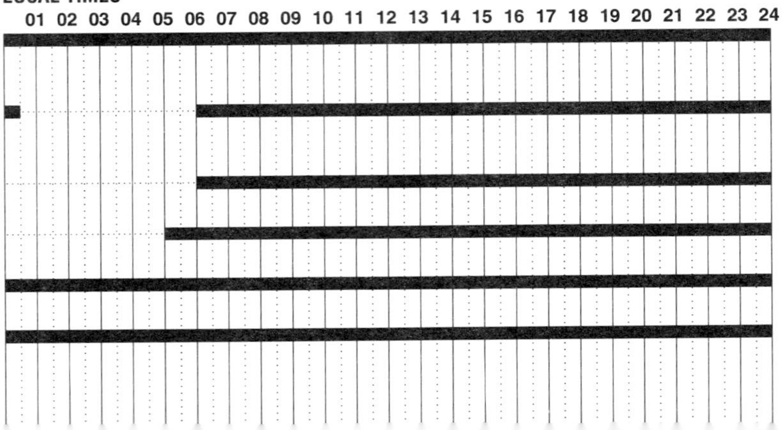

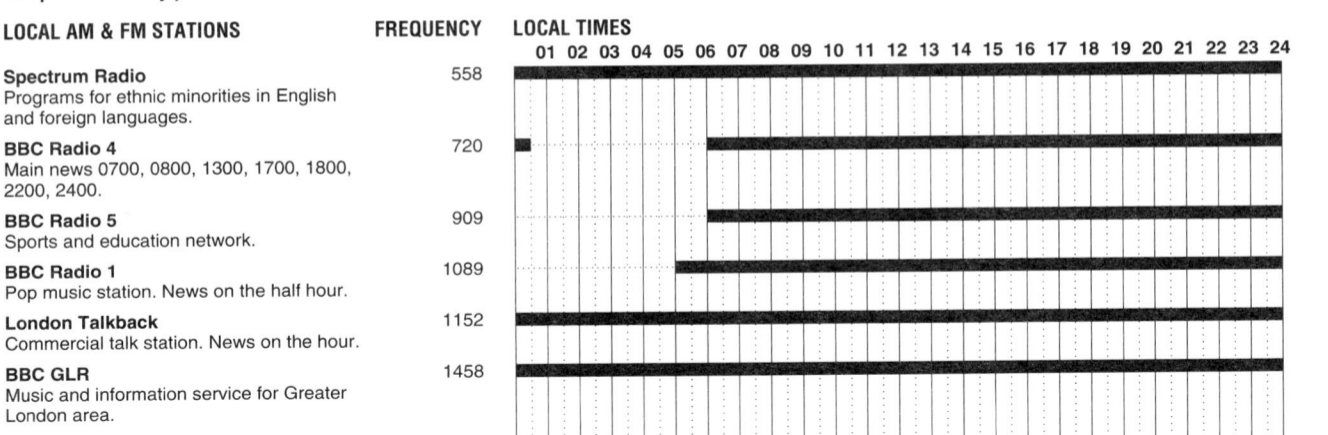

LONDON 105

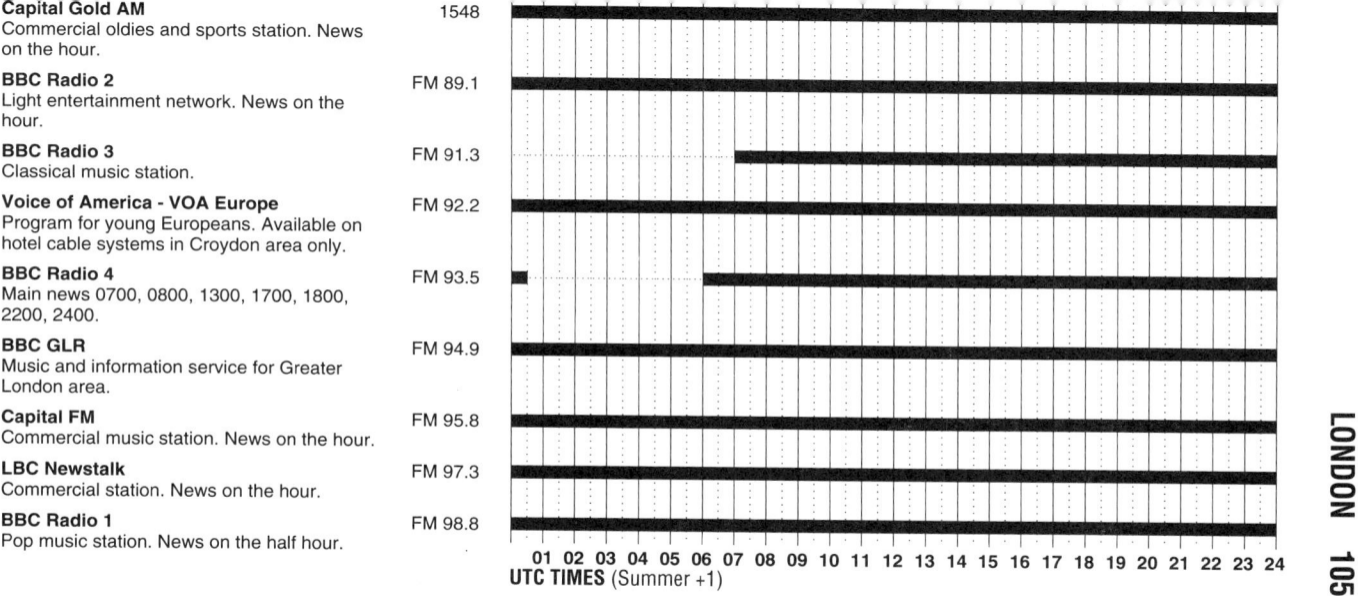

Station	Frequency	Description
Capital Gold AM	1548	Commercial oldies and sports station. News on the hour.
BBC Radio 2	FM 89.1	Light entertainment network. News on the hour.
BBC Radio 3	FM 91.3	Classical music station.
Voice of America - VOA Europe	FM 92.2	Program for young Europeans. Available on hotel cable systems in Croydon area only.
BBC Radio 4	FM 93.5	Main news 0700, 0800, 1300, 1700, 1800, 2200, 2400.
BBC GLR	FM 94.9	Music and information service for Greater London area.
Capital FM	FM 95.8	Commercial music station. News on the hour.
LBC Newstalk	FM 97.3	Commercial station. News on the hour.
BBC Radio 1	FM 98.8	Pop music station. News on the half hour.

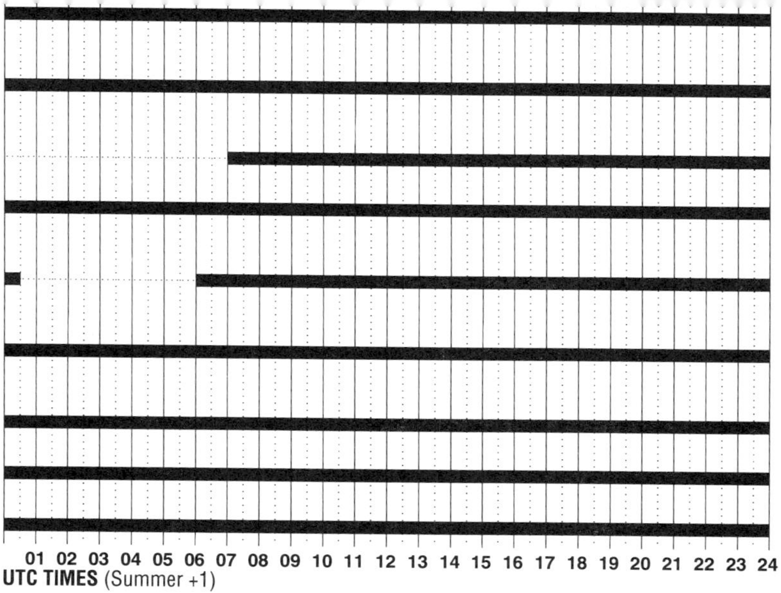

01 02 03 04 05 06 07 08 09 10 11 12 13 14 15 16 17 18 19 20 21 22 23 24
UTC TIMES (Summer +1)

106 LONDON

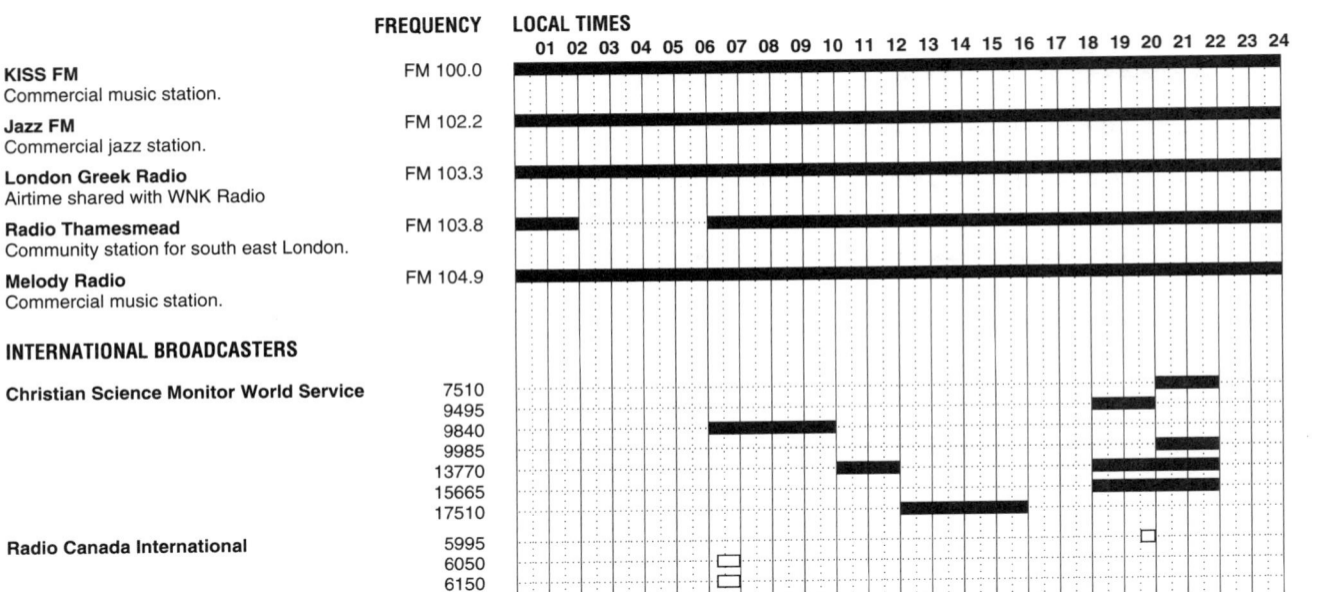

FREQUENCY		LOCAL TIMES

KISS FM — FM 100.0
Commercial music station.

Jazz FM — FM 102.2
Commercial jazz station.

London Greek Radio — FM 103.3
Airtime shared with WNK Radio

Radio Thamesmead — FM 103.8
Community station for south east London.

Melody Radio — FM 104.9
Commercial music station.

INTERNATIONAL BROADCASTERS

Christian Science Monitor World Service
7510
9495
9840
9985
13770
15665
17510

Radio Canada International
5995
6050
6150

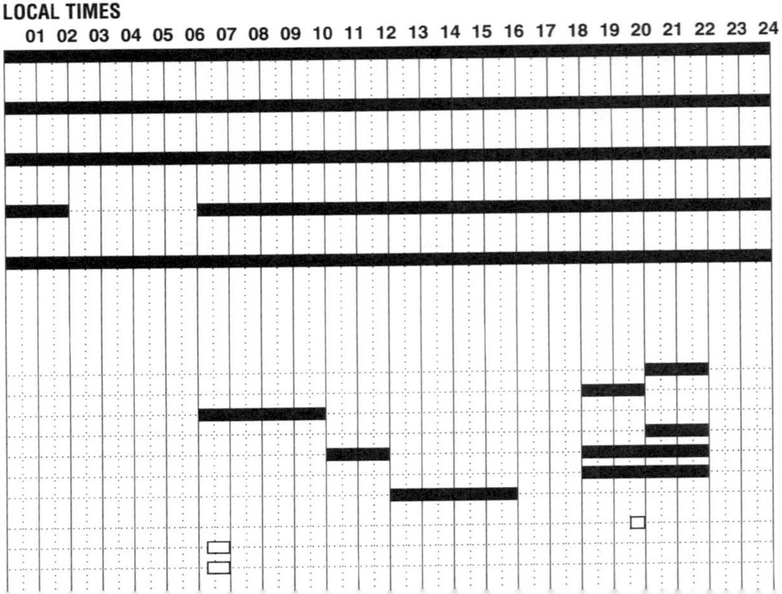

LONDON 107

Radio France International

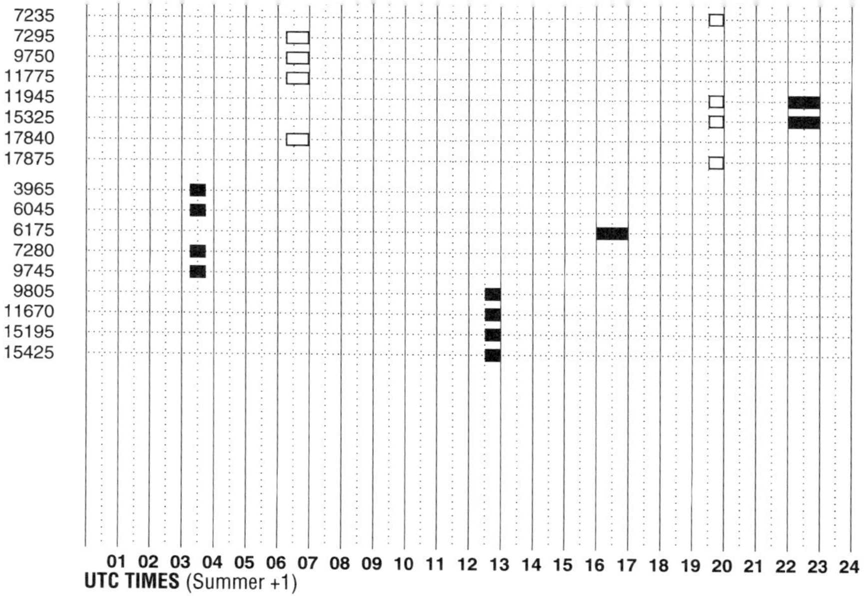

Frequency	Schedule
7235	
7295	
9750	
11775	
11945	
15325	
17840	
17875	
3965	
6045	
6175	
7280	
9745	
9805	
11670	
15195	
15425	

UTC TIMES (Summer +1)

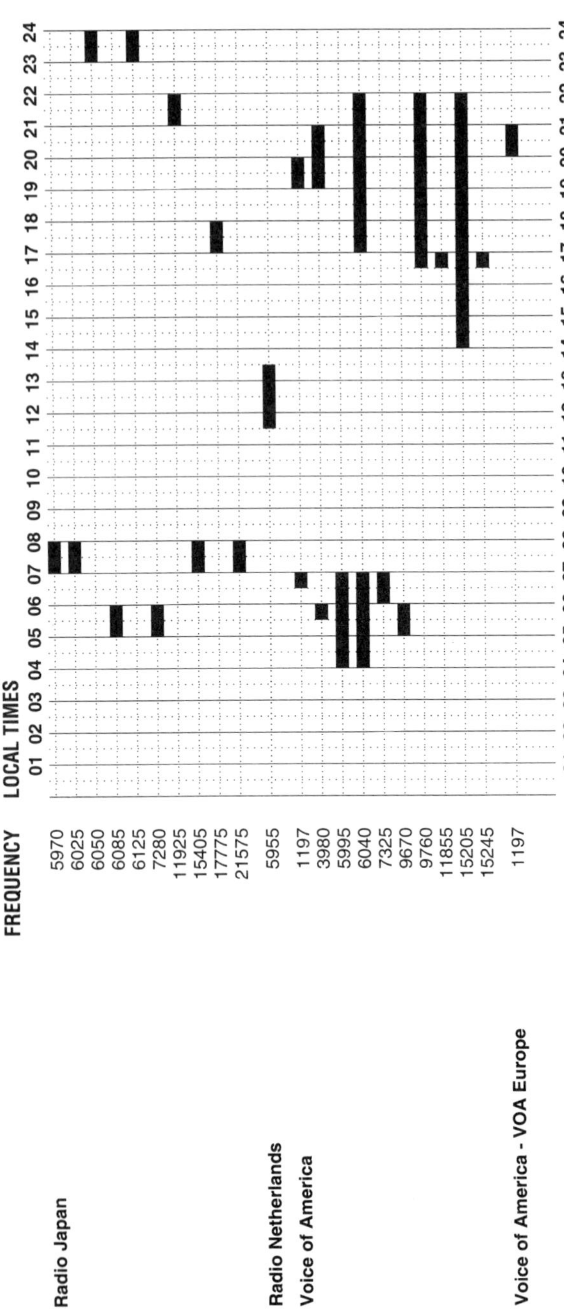

108 LONDON

Broadcaster	Frequency
Radio Japan	5970, 6025, 6050, 6085, 6125, 7280, 11925, 15405, 17775, 21575
Radio Netherlands	5955
Voice of America	1197, 3980, 5995, 6040, 7325, 9670, 9760, 11855, 15205, 15245
Voice of America - VOA Europe	1197

LOS ANGELES 109

LOS ANGELES, USA
Hours difference from UTC (GMT): -8 (Summer -7)
Main business languages: English, Spanish
Electricity: 60Hz, 110V
Currency: US Dollar
Telephone country prefix and area code: 1 (213)

LOCAL AM & FM STATIONS	FREQUENCY
KABC Newstalk.	790
KFWB All news.	980
KNX All news.	1070
KUSC Fine arts (non-commercial station).	FM 91.5
KKGO-FM Fine arts.	FM 105.1

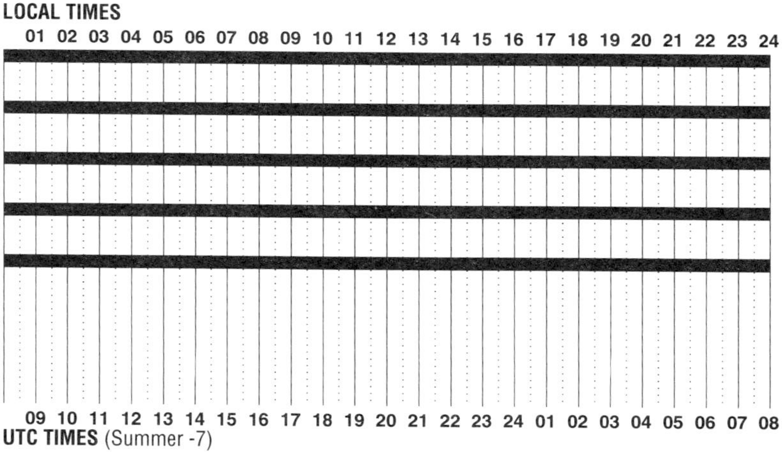

LOCAL TIMES
01 02 03 04 05 06 07 08 09 10 11 12 13 14 15 16 17 18 19 20 21 22 23 24
09 10 11 12 13 14 15 16 17 18 19 20 21 22 23 24 01 02 03 04 05 06 07 08
UTC TIMES (Summer -7)

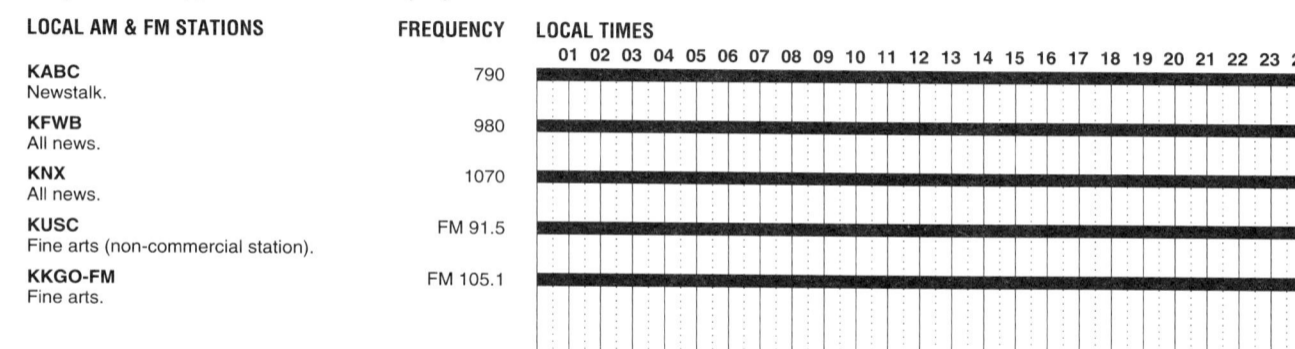

110 LOS ANGELES

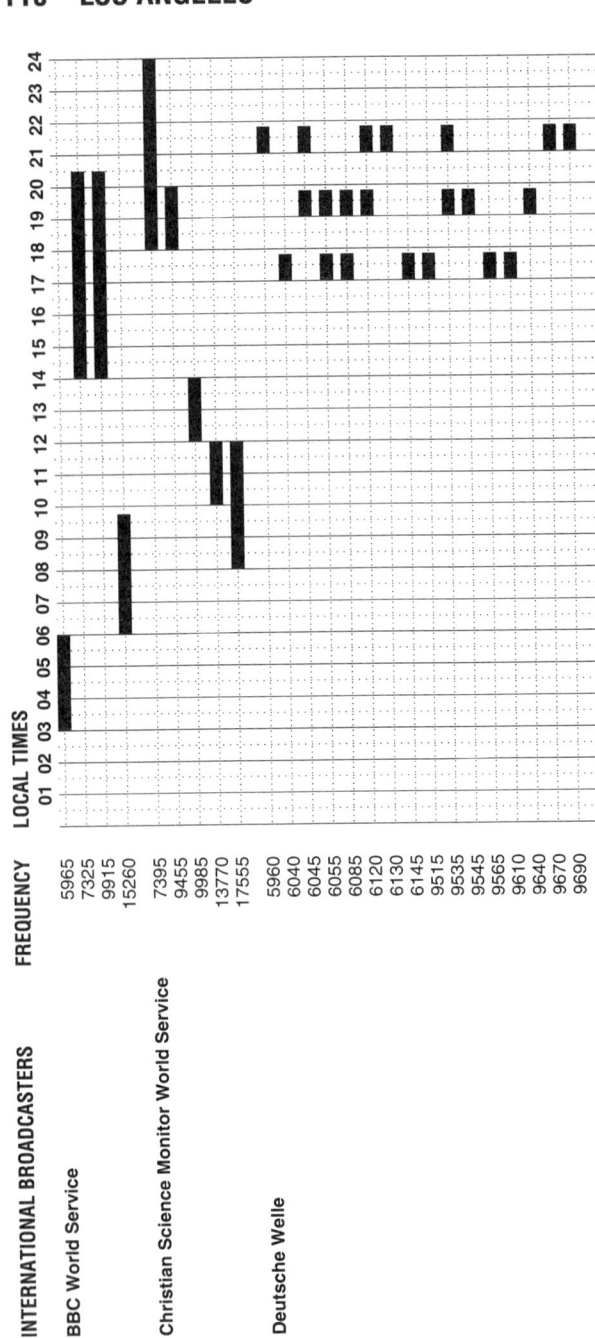

INTERNATIONAL BROADCASTERS	FREQUENCY
BBC World Service	5965
	7325
	9915
	15260
Christian Science Monitor World Service	7395
	9455
	9985
	13770
	17555
Deutsche Welle	5960
	6040
	6045
	6055
	6085
	6120
	6130
	6145
	9515
	9535
	9545
	9565
	9610
	9640
	9670
	9690

LOS ANGELES 111

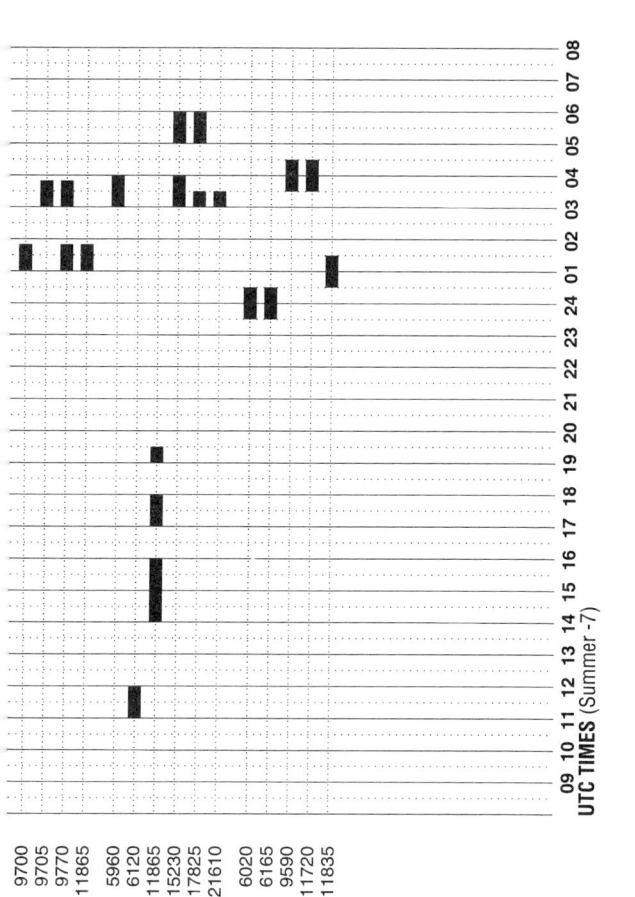

Radio Japan

Frequency	Schedule
9700	
9705	
9770	
11865	
5960	
6120	
11865	
15230	
17825	
21610	

Radio Netherlands

Frequency	Schedule
6020	
6165	
9590	
11720	
11835	

UTC TIMES (Summer -7): 09 10 11 12 13 14 15 16 17 18 19 20 21 22 23 24 01 02 03 04 05 06 07 08

112 MADRID

MADRID, SPAIN
Hours difference from UTC (GMT): +1 (Summer +2)
Main business languages: Spanish, English
Electricity: 50Hz, voltage varies
Currency: Peseta
Telephone country prefix and area code: 34 (1)

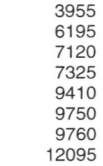

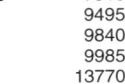

INTERNATIONAL BROADCASTERS	FREQUENCY
BBC World Service	3955
	6195
	7120
	7325
	9410
	9750
	9760
	12095
	15070
	15575
	17640
	17705
Christian Science Monitor World Service	7510
	9495
	9840
	9985
	13770
	15665

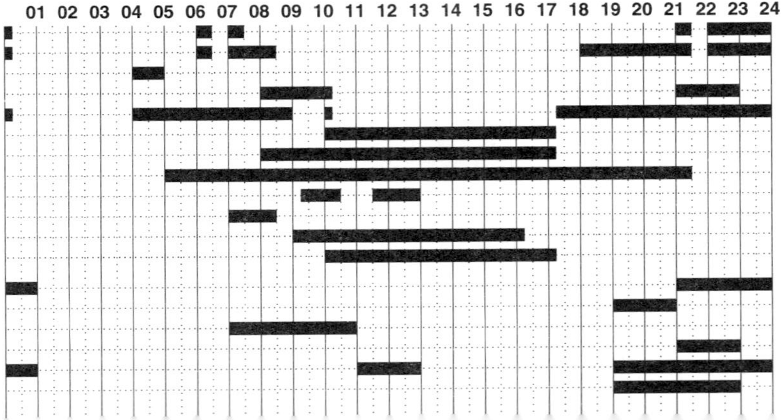

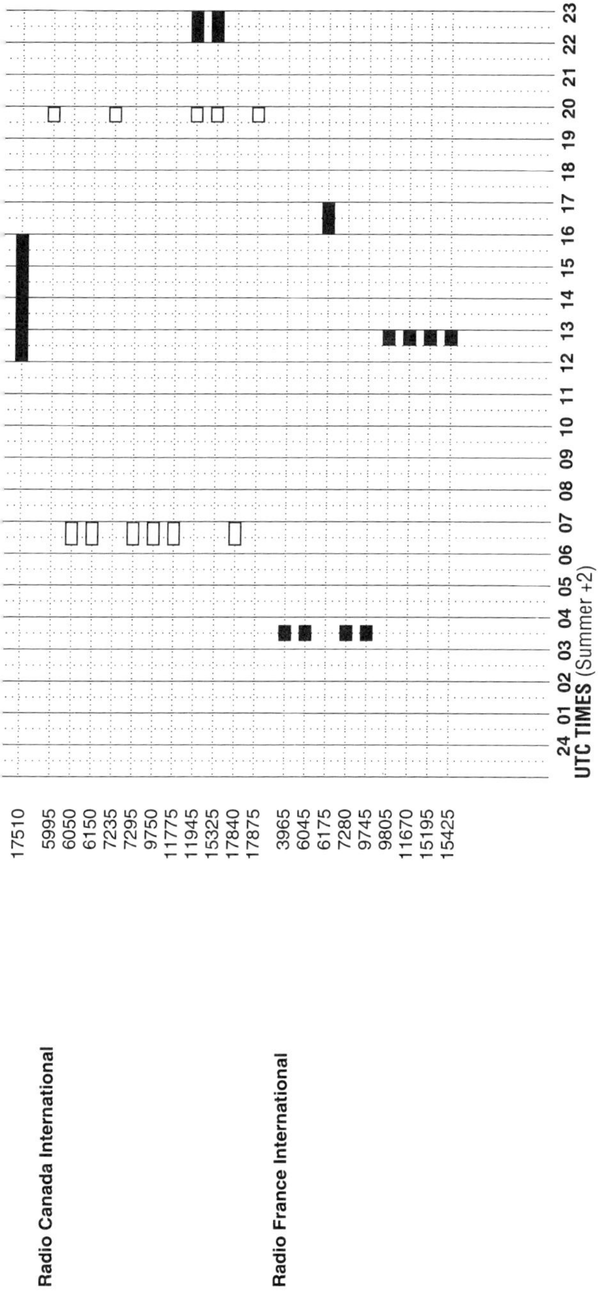

MADRID 113

Radio Canada International

17510
5995
6050
6150
7235
7295
9750
11775
11945
15325
17840
17875

Radio France International

3965
6045
6175
7280
9745
9805
11670
15195
15425

UTC TIMES (Summer +2)

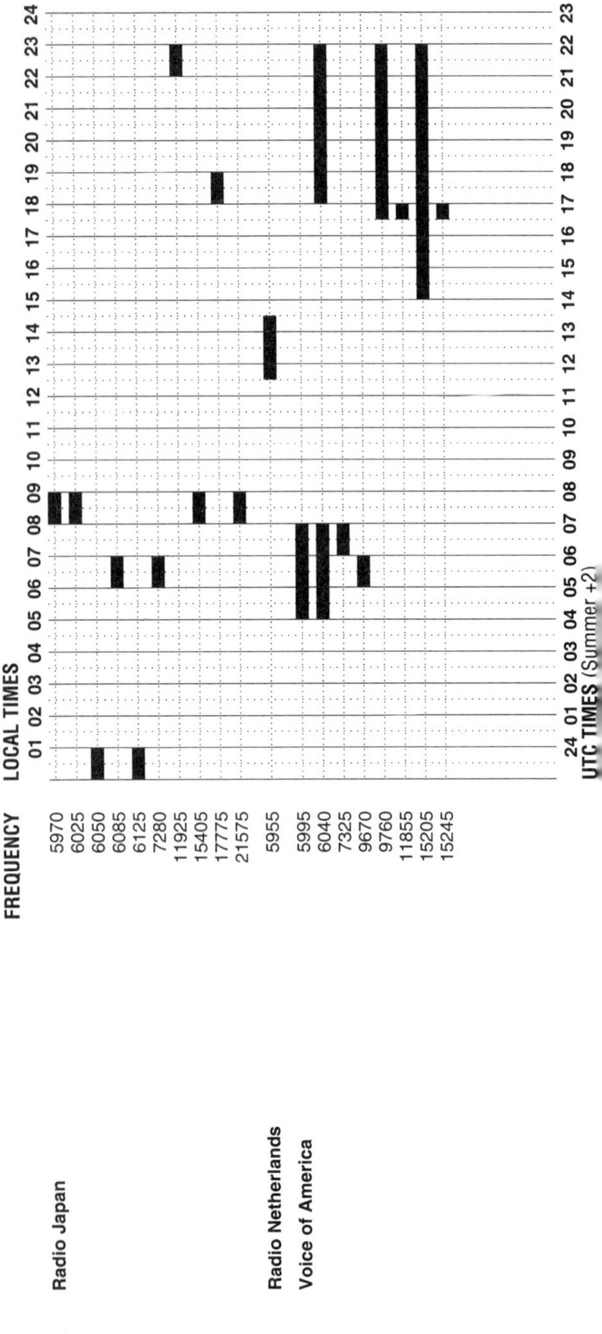

114 MADRID

	FREQUENCY	
Radio Japan	5970, 6025, 6050, 6085, 6125, 7280, 11925, 15405, 17775, 21575	
Radio Netherlands	5955	
Voice of America	5995, 6040, 7325, 9670, 9760, 11855, 15205, 15245	

MANILA, PHILIPPINES

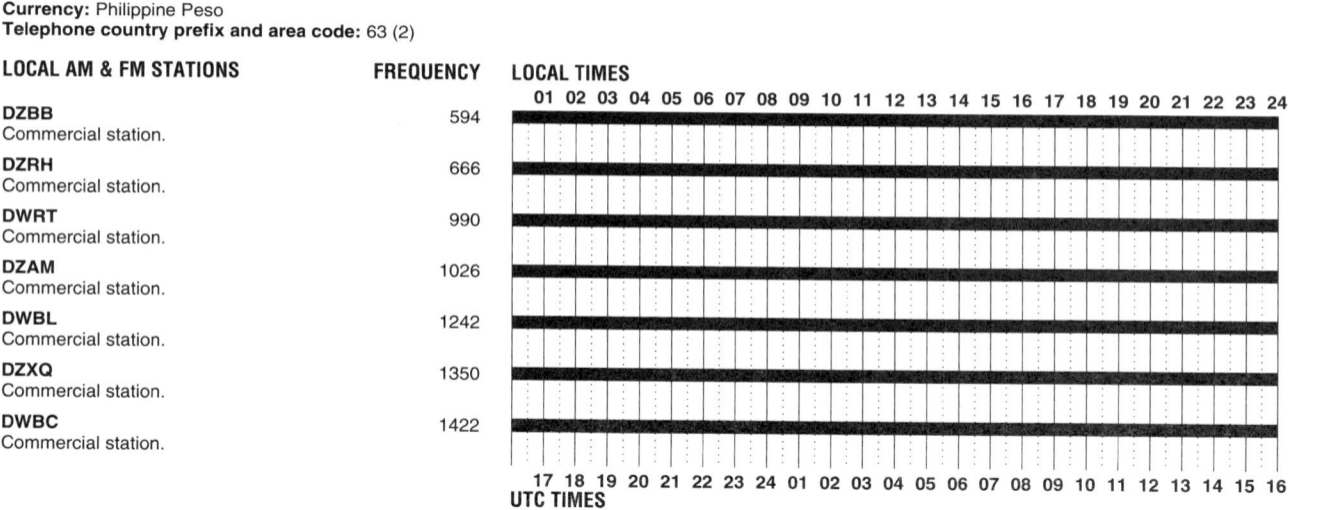

Hours difference from UTC (GMT): +8
Main business languages: English
Electricity: 60Hz, 110/220V
Currency: Philippine Peso
Telephone country prefix and area code: 63 (2)

MANILA 115

LOCAL AM & FM STATIONS	FREQUENCY
DZBB Commercial station.	594
DZRH Commercial station.	666
DWRT Commercial station.	990
DZAM Commercial station.	1026
DWBL Commercial station.	1242
DZXQ Commercial station.	1350
DWBC Commercial station.	1422

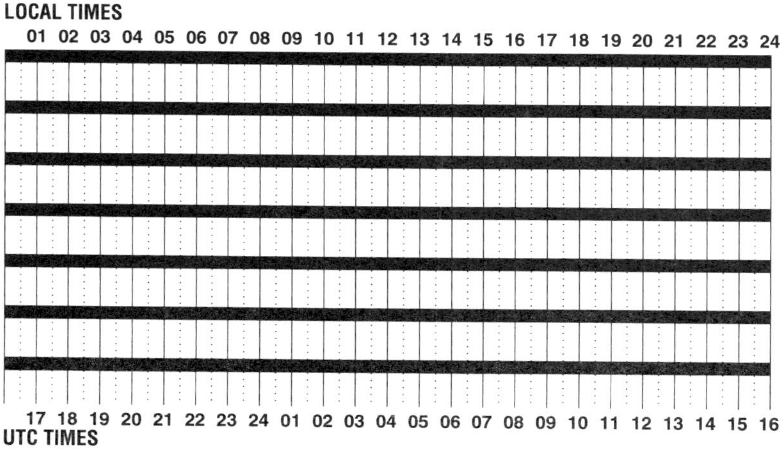

116 MANILA

	FREQUENCY	LOCAL TIMES 01–24
DWCT Commercial station.	FM 88.3	
DWTM	FM 89.9	
DWKY	FM 91.5	
DWRX Commercial station.	FM 93.1	
DWKC Commercial station.	FM 93.9	
DWLL Commercial station.	FM 94.7	
DWRK Commercial station.	FM 96.3	
DWAD Commercial station.	FM 97.9	
DYFR Religious station.	FM 98.7	
DWRT Commercial station.	FM 99.5	
DWKS Commercial station.	FM 101.1	

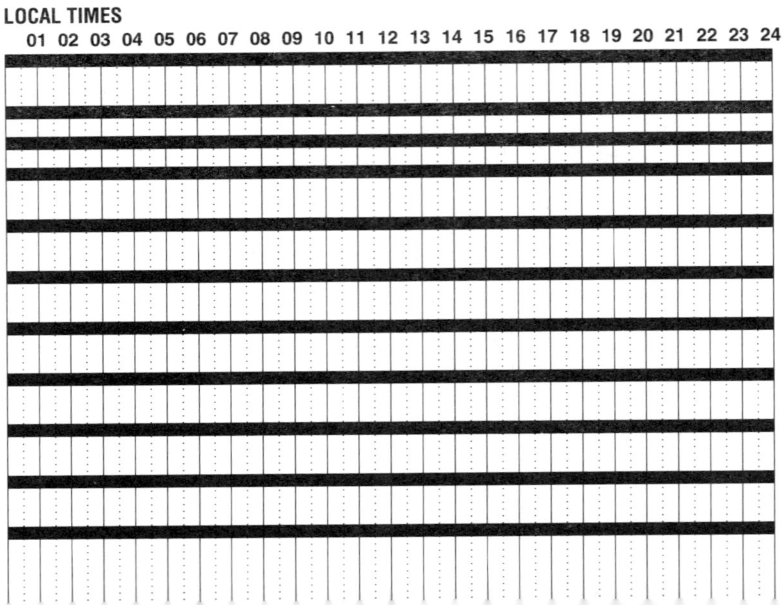

MANILA 117

DWSM
Commercial station.

FM 102.7

INTERNATIONAL BROADCASTERS

BBC World Service

3915	
5975	
6195	
9570	
9740	
11750	
11955	
15310	
15340	
15360	
17790	
17830	
21715	

Christian Science Monitor World Service

13625	
15610	
17555	

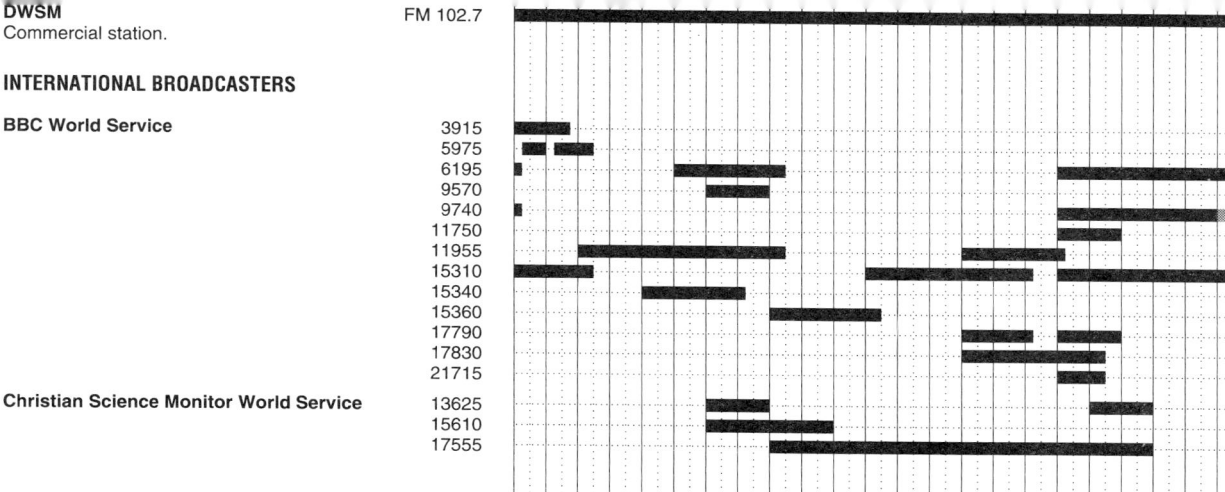

17 18 19 20 21 22 23 24 01 02 03 04 05 06 07 08 09 10 11 12 13 14 15 16
UTC TIMES

118 MANILA

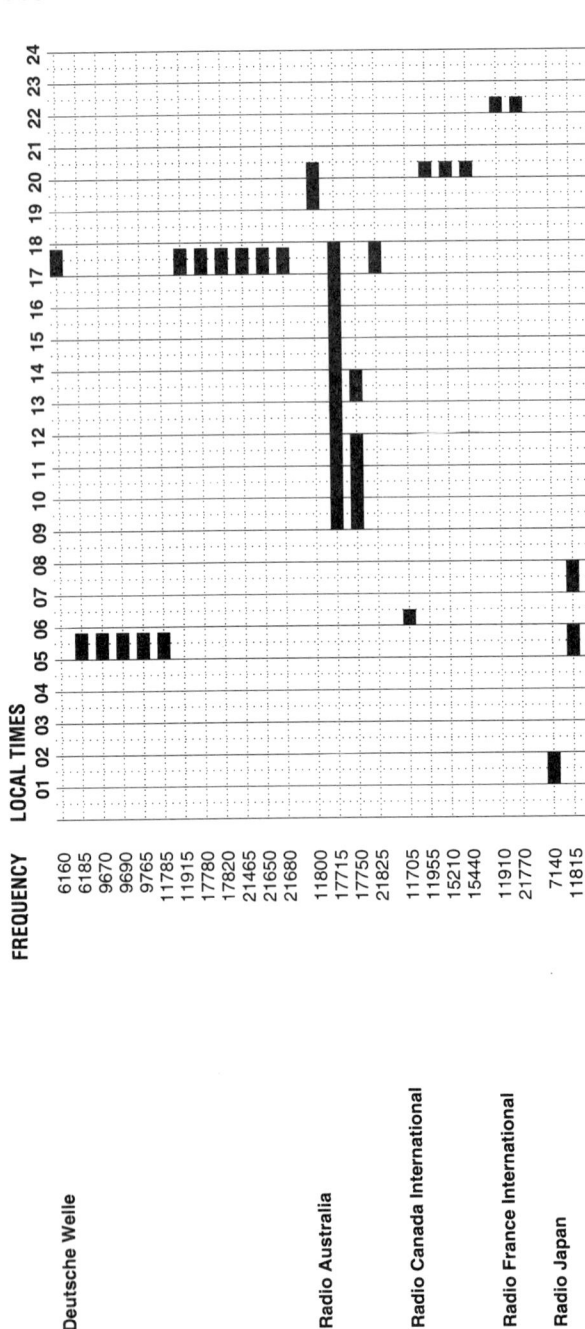

Station	Frequency	Local Times
Deutsche Welle	6160	17
	6185	05–06
	9670	05–06
	9690	05–06
	9765	05–06
	11785	05–06
	11915	17–18
	17780	17–18
	17820	17–18
	21465	17–18
	21650	17–18
	21680	17–18
Radio Australia	11800	19–20
	17715	09–11, 12–13, 17–18
	17750	13–14
	21825	17–18
Radio Canada International	11705	06
	11955	20
	15210	20
	15440	20
Radio France International	11910	22–23
	21770	22–23
Radio Japan	7140	01–02
	11815	05–06, 07–08

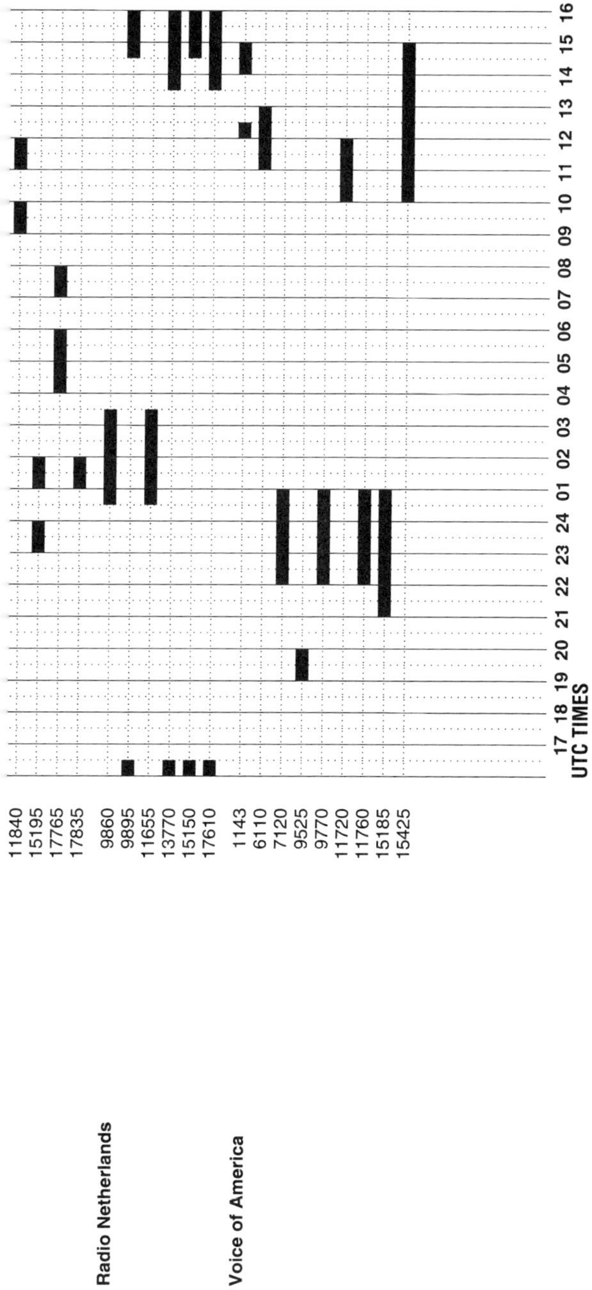

MANILA 119

Radio Netherlands

Frequency
11840
15195
17765
17835

Voice of America

Frequency
9860
9895
11655
13770
15150
17610
1143
6110
7120
9525
9770
11720
11760
15185
15425

120 MELBOURNE

MELBOURNE, AUSTRALIA
Hours difference from UTC (GMT): +10 (Summer +11)
Main business languages: English
Electricity: 50Hz, 240/415V
Currency: Australian Dollars
Telephone country prefix and area code: 61 (3)

LOCAL AM & FM STATIONS	FREQUENCY
ABC Radio National Non-commercial station.	621
ABC Metropolitan Service Regional non-commercial station.	774
3AW Commercial station. News on the hour.	1278
3KKZ-FM Commercial station. News on the hour.	FM 104.3
3MMM-FM Commercial station. News on the hour.	FM 105.1
ABC FM Radio Service Non-commercial station.	FM 105.9

MELBOURNE 121

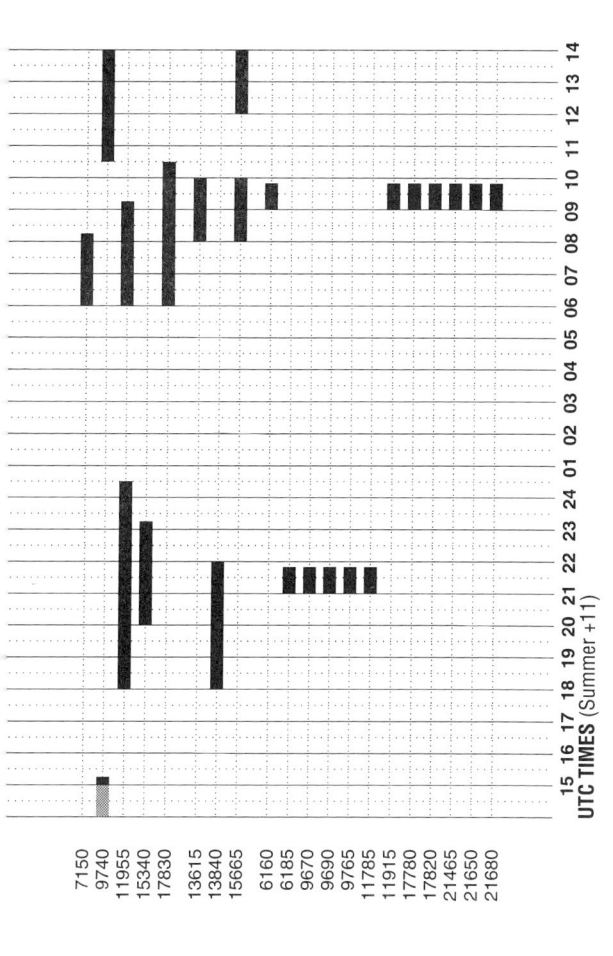

INTERNATIONAL BROADCASTERS

BBC World Service

Frequency	Schedule
7150	
9740	12–14
11955	07–09, 22–24
15340	07–09, 23–24
17830	07–10

Christian Science Monitor World Service

Frequency	Schedule
13615	09–10
13840	12–13
15665	18–22

Deutsche Welle

Frequency	Schedule
6160	
6185	
9670	19–21
9690	
9765	
11785	
11915	
17780	09–10
17820	
21465	
21650	
21680	

UTC TIMES (Summer +11)

122 MELBOURNE

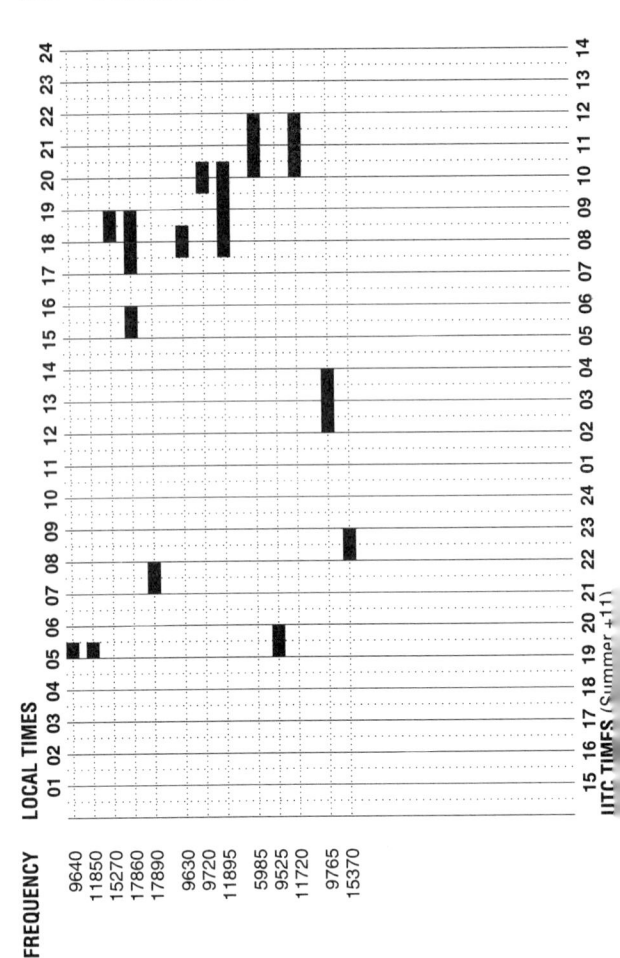

FREQUENCY		
Radio Japan	9640 11850 15270 17860 17890	
Radio Netherlands	9630 9720 11895	
Voice of America	5985 9525 11720	
Voice of Free China	9765 15370	

LOCAL TIMES

UTC TIMES (Summer +11)

MEXICO CITY, MEXICO

Hours difference from UTC (GMT): -6
Main business languages: Spanish, English
Electricity: 60Hz, 110/220V
Currency: Mexican Peso
Telephone country prefix and area code: 52 (5)

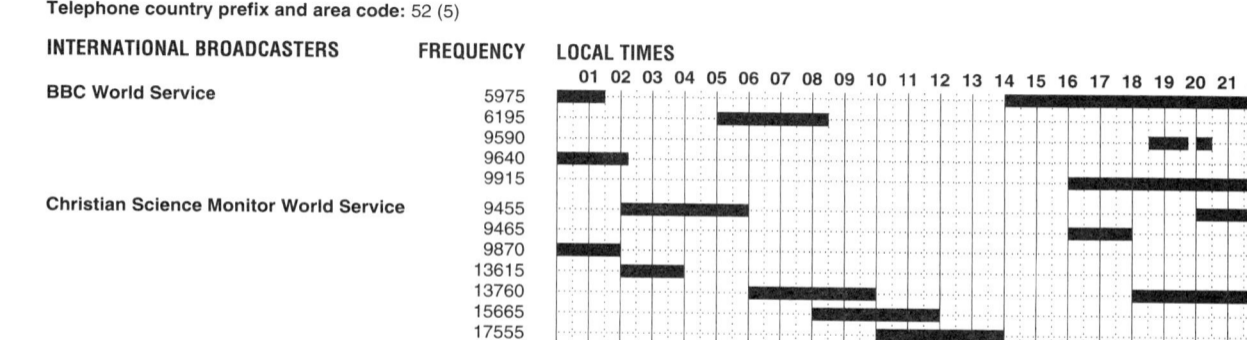

INTERNATIONAL BROADCASTERS	FREQUENCY
BBC World Service	5975
	6195
	9590
	9640
	9915
Christian Science Monitor World Service	9455
	9465
	9870
	13615
	13760
	15665
	17555

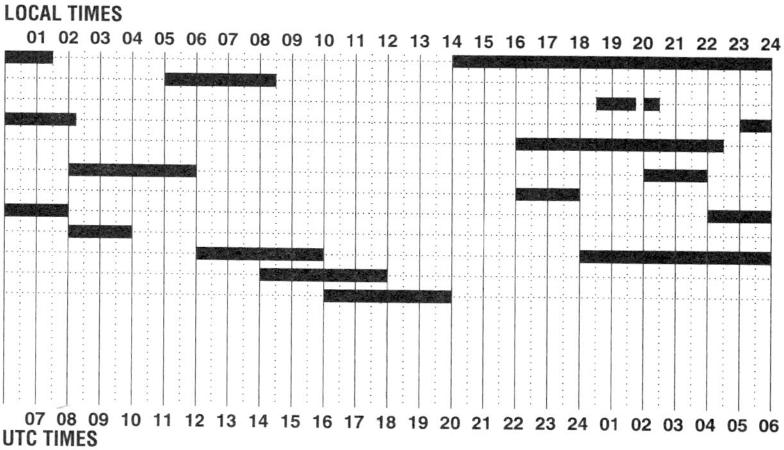

MEXICO CITY 123

124 MEXICO CITY

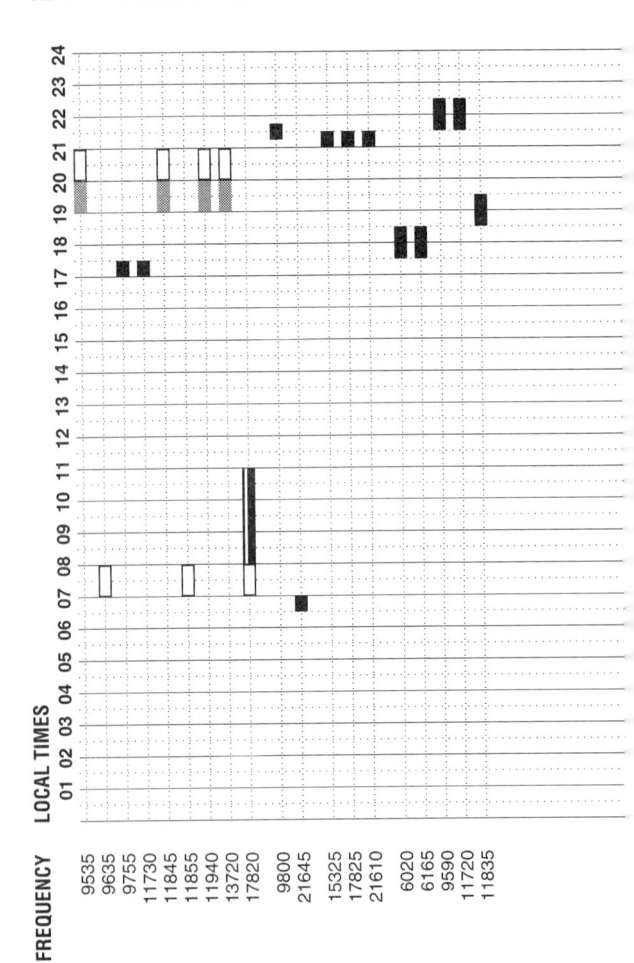

Station	Frequency	Local Times
Radio Canada International	9535	19–21
	9635	19–21
	9755	17
	11730	17
	11845	19–21
	11855	19–21
	11940	07
	13720	07
	17820	07–11
Radio France International	9800	21–22
	21645	06
Radio Japan	15325	21–22
	17825	21–22
	21610	21–22
Radio Netherlands	6020	18
	6165	18
	9590	19
	11720	22
	11835	22

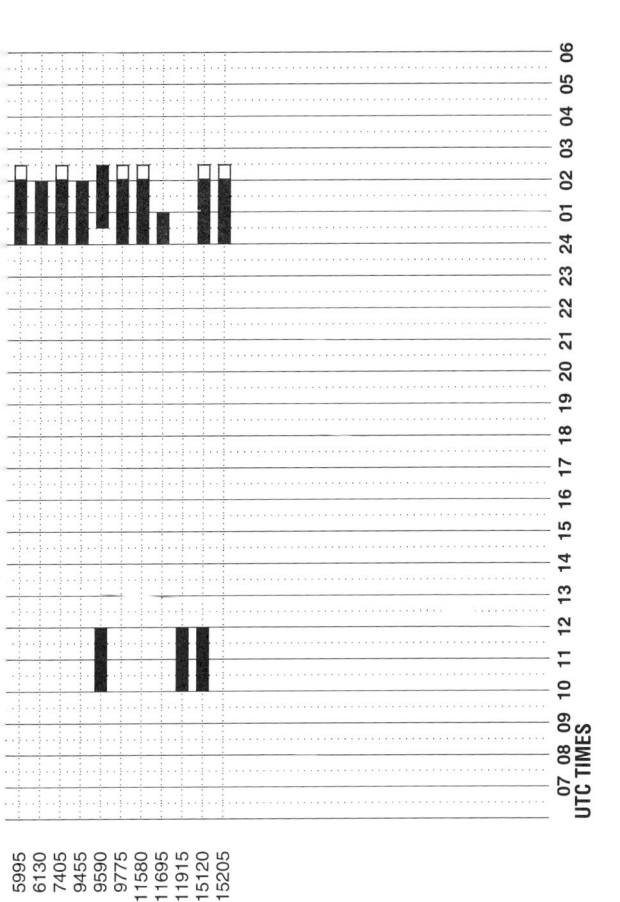

MEXICO CITY 125

Voice of America

Frequencies (kHz): 5995, 6130, 7405, 9455, 9590, 9775, 11580, 11695, 11915, 15120, 15205

UTC TIMES

126 MILAN

MILAN, ITALY
Hours difference from UTC (GMT): +1 (Summer +2)
Main business languages: Italian, English
Electricity: 50Hz, voltage varies
Currency: Italian Lira
Telephone country prefix and area code: 39 (2)

LOCAL AM & FM STATIONS | FREQUENCY

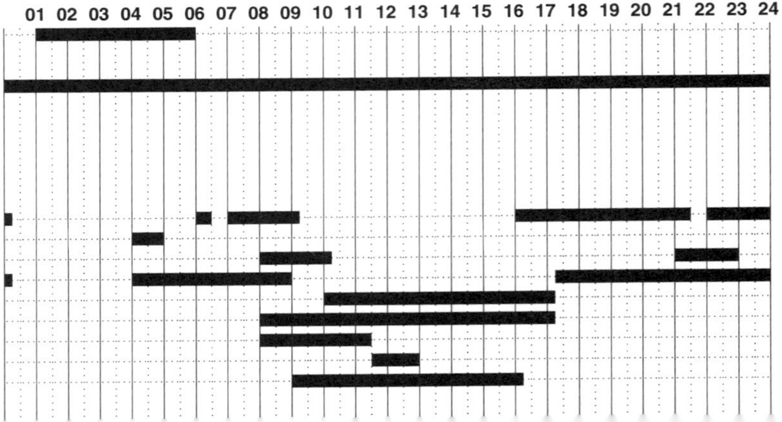

RAI Night Program — 900
News summary at 3 minutes past each hour.

Voice of America - VOA Europe — FM 96.2
Program for young Europeans. News on the hour.

INTERNATIONAL BROADCASTERS

BBC World Service
6195
7120
7325
9410
9750
9760
12095
15070
17640

MILAN

Christian Science Monitor World Service

7510
9495
9840
9985
13770
15665
17510

Radio Canada International

5995
6050
6150
7235
7295
9750
11775
11945
15325
17840
17875

UTC TIMES (Summer +2)

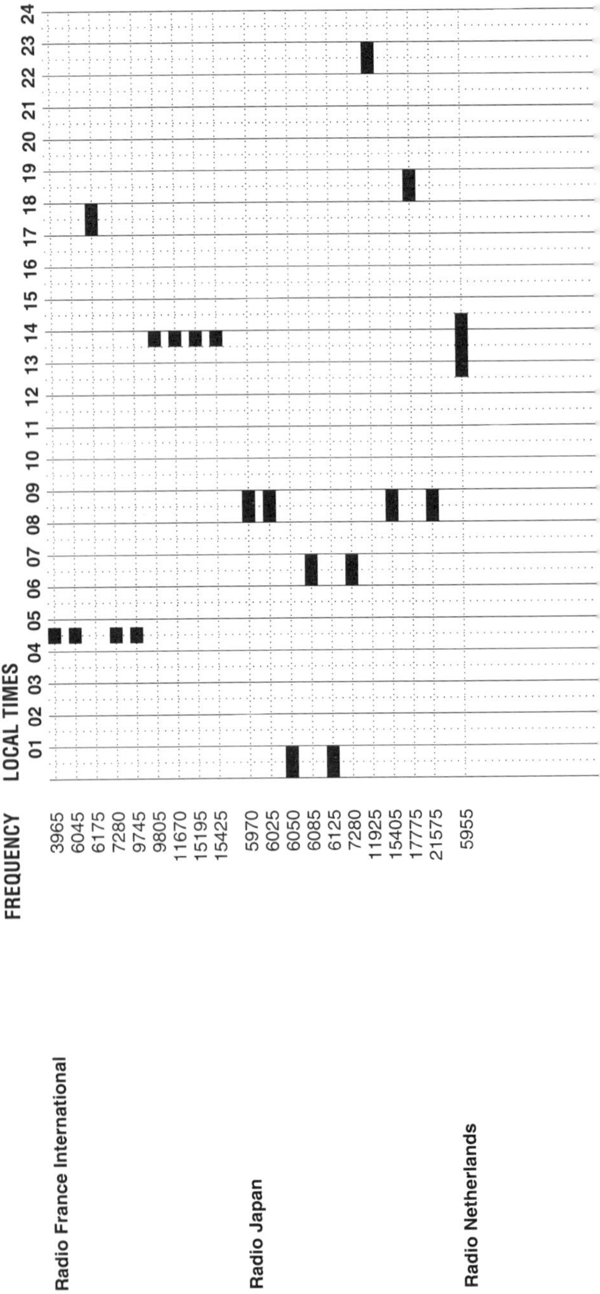

128 MILAN

MILAN 129

Voice of America

Frequency	Broadcast times (UTC)
5995	
6040	
7325	
9670	
9760	
11855	
15205	
15245	

UTC TIMES (Summer +2)

MONTREAL

MONTREAL, CANADA
Hours difference from UTC (GMT): -5 (Summer -4)
Main business languages: French, English
Electricity: 60Hz, 115/230V
Currency: Canadian Dollar
Telephone country prefix and area code: 1 (514)

LOCAL AM & FM STATIONS	FREQUENCY
CFCF Commercial station.	600
CKO Commercial all-news station.	650
CJAD Commercial station.	800
CBM Canadian Broadcasting Corporation. Non-commercial station.	940
CHTX Commercial station.	980
CFMB Commercial station.	1410 FM 90.3
CFQR Commercial easy listening station.	FM 92.5

LOCAL TIMES
01 02 03 04 05 06 07 08 09 10 11 12 13 14 15 16 17 18 19 20 21 22 23 24

MONTREAL 131

CBM-FM
Canadian Broadcasting Corporation.
Non-commercial station.

FM 93.5

INTERNATIONAL BROADCASTERS

BBC World Service

6175
9515
9590
9915
15220
17840

Christian Science Monitor World Service

7395
9985
13770
15665

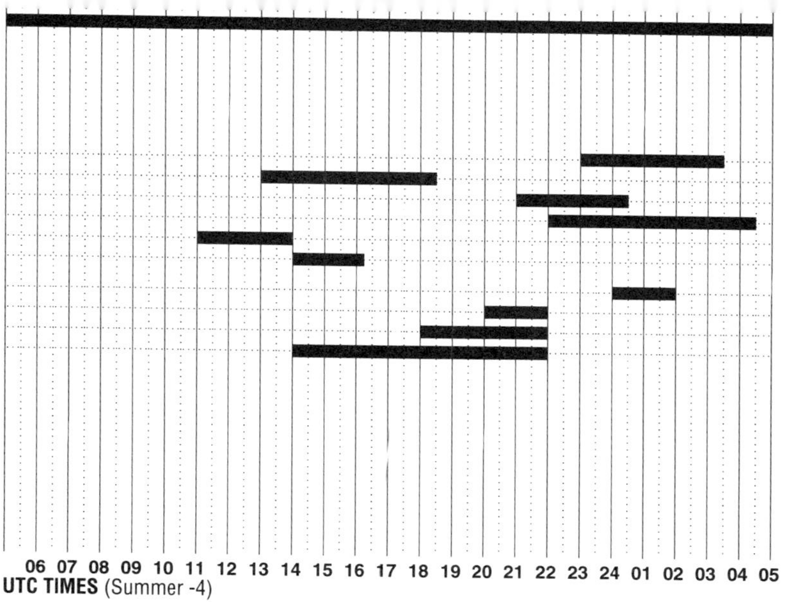

UTC TIMES (Summer −4)

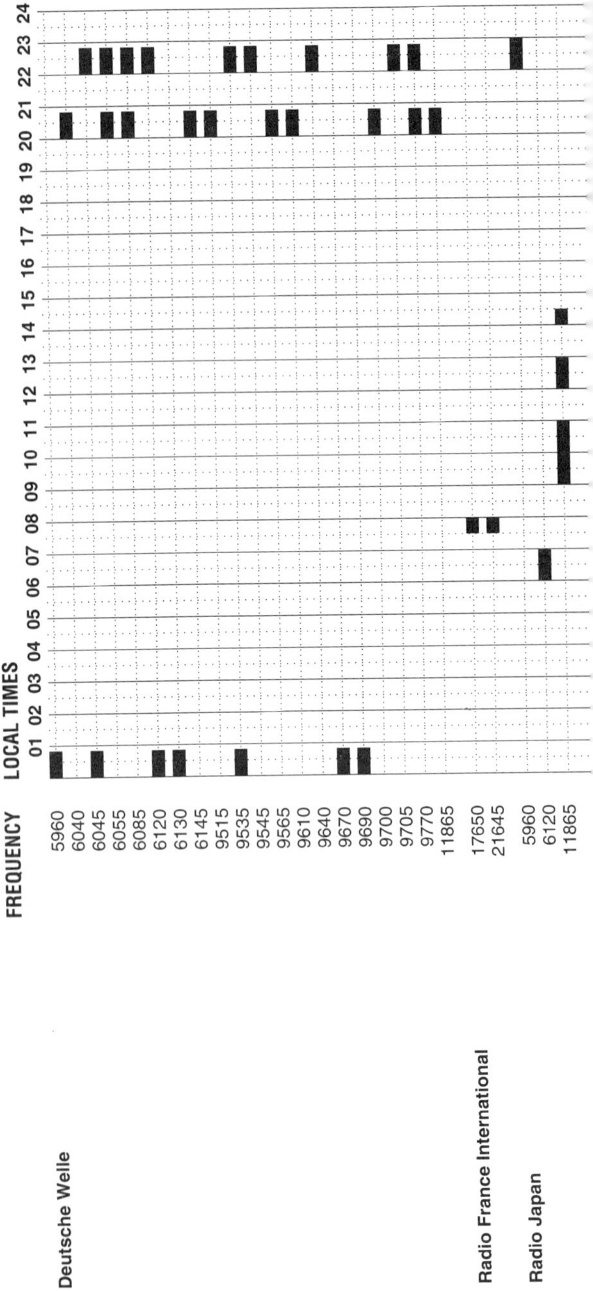

132 MONTREAL

Deutsche Welle

Frequency	Local Times
5960	20–21
6040	00–01
6045	20–21, 22–23
6055	22–23
6085	22–23
6120	00–01
6130	00–01
6145	20–21
9515	00–01
9535	22–23
9545	20–21
9565	22–23
9610	20–21
9640	00–01
9670	00–01
9690	20–21
9700	22–23
9705	22–23
9770	20–21
11865	22–23

Radio France International

Frequency	Local Times
17650	07–08
21645	07–08

Radio Japan

Frequency	Local Times
5960	06–07
6120	—
11865	09–10, 12–13, 14–15

MONTREAL 133

Radio Netherlands

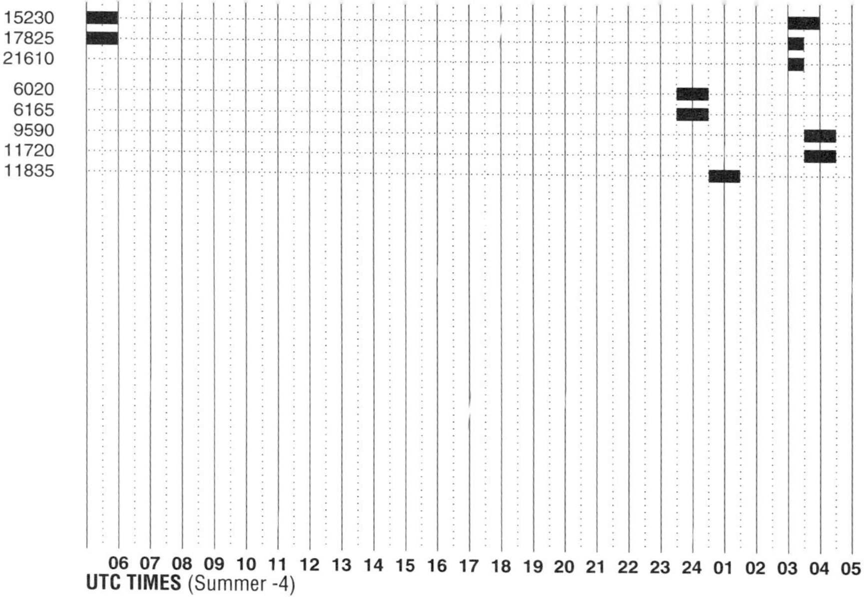

15230
17825
21610
6020
6165
9590
11720
11835

UTC TIMES (Summer -4)

134 MOSCOW

MOSCOW, RUSSIA
Hours difference from UTC (GMT): +3 (Summer +4)
Main business languages: Russian, English
Electricity: 50Hz, 127/220V
Currency: Rouble
Telephone country prefix and area code: 7 (095)

LOCAL AM & FM STATIONS — FREQUENCY

Voice of America - VOA Europe — 918
Program for young Europeans. News on the hour.

INTERNATIONAL BROADCASTERS

BBC World Service
3955
6180
6195
9410
15070
17640

Christian Science Monitor World Service
7510
9495
9840
9985
13770

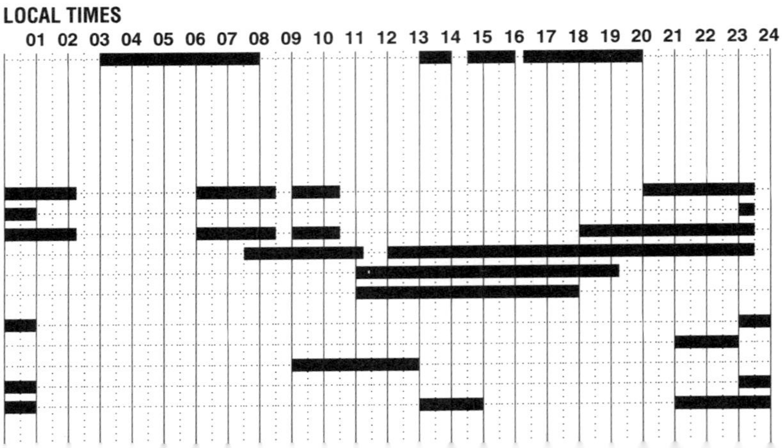

MOSCOW 135

Radio Canada International

15665	
17510	
5995	
7235	
9555	
11915	
11935	
13650	
15315	
15325	
17820	
21545	

Radio France International

9805	
11670	
15195	
15425	

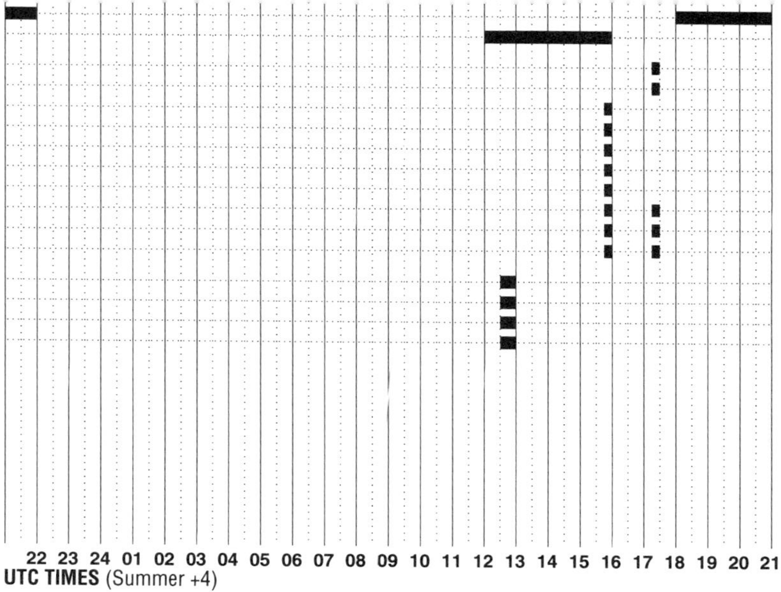

UTC TIMES (Summer +4)

136 MOSCOW

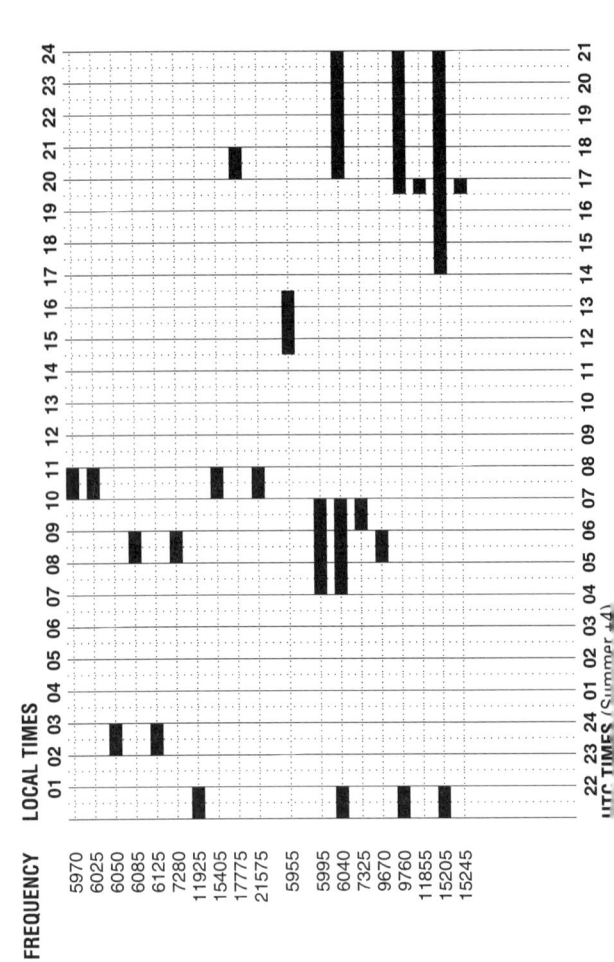

FREQUENCY		
	Radio Japan	5970, 6025, 6050, 6085, 6125, 7280, 11925, 15405, 17775, 21575
	Radio Netherlands	5955
	Voice of America	5995, 6040, 7325, 9670, 9760, 11855, 15205, 15245

MUNICH 137

MUNICH, GERMANY
Hours difference from UTC (GMT): +1 (Summer +2)
Main business languages: German, English
Electricity: 50Hz, 220V
Currency: Deutschmark
Telephone country prefix and area code: 49 (89)

LOCAL AM & FM STATIONS

	FREQUENCY
American Forces Network News on the hour.	1107
Deutschlandfunk Program for listeners abroad.	1269

138 MUNICH

INTERNATIONAL BROADCASTERS	FREQUENCY
BBC World Service	198
	648
	1296
	6180
	6195
	7120
	7325
	9410
	9660
	9750
	9760
	12095
	15070
	15575
	17640
Christian Science Monitor World Service	7510
	9495
	9840
	9985
	13770
	15665
	17510
Radio Canada International	5995
	6050
	6150

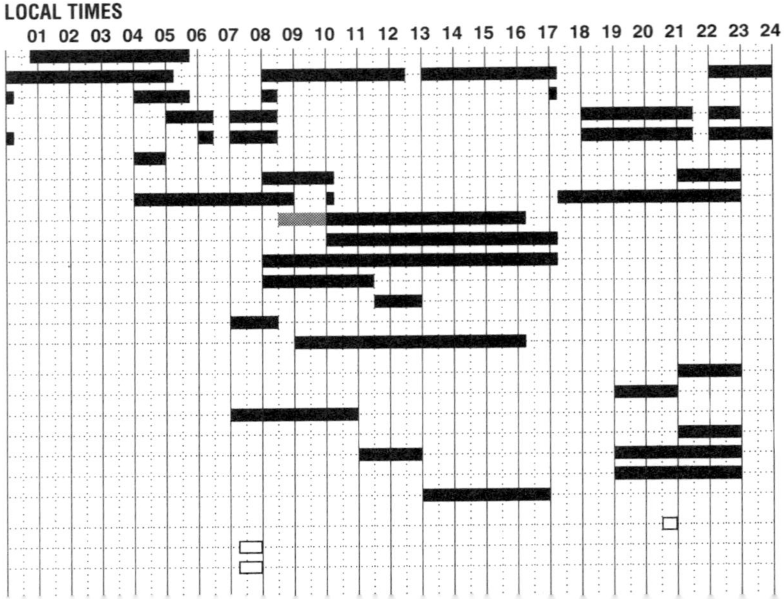

MUNICH 139

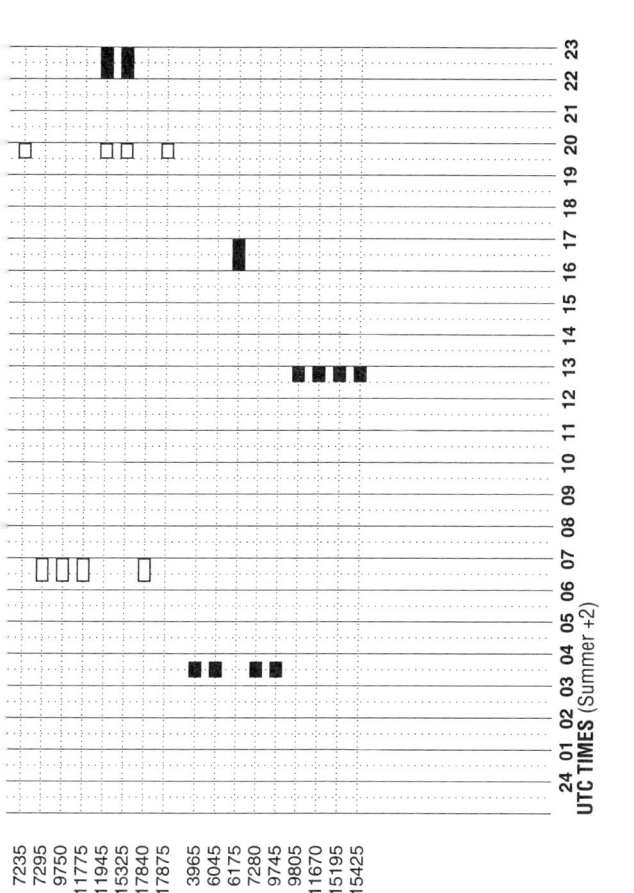

Radio France International

Frequency	Times (UTC)
7235	19-20
7295	
9750	19-20
11775	
11945	19-20
15325	21-22
17840	
17875	
3965	06-07
6045	06-07
6175	02-03
7280	06-07
9745	03-04
9805	
11670	12-13
15195	16-17
15425	12-13

UTC TIMES (Summer +2)

140 MUNICH

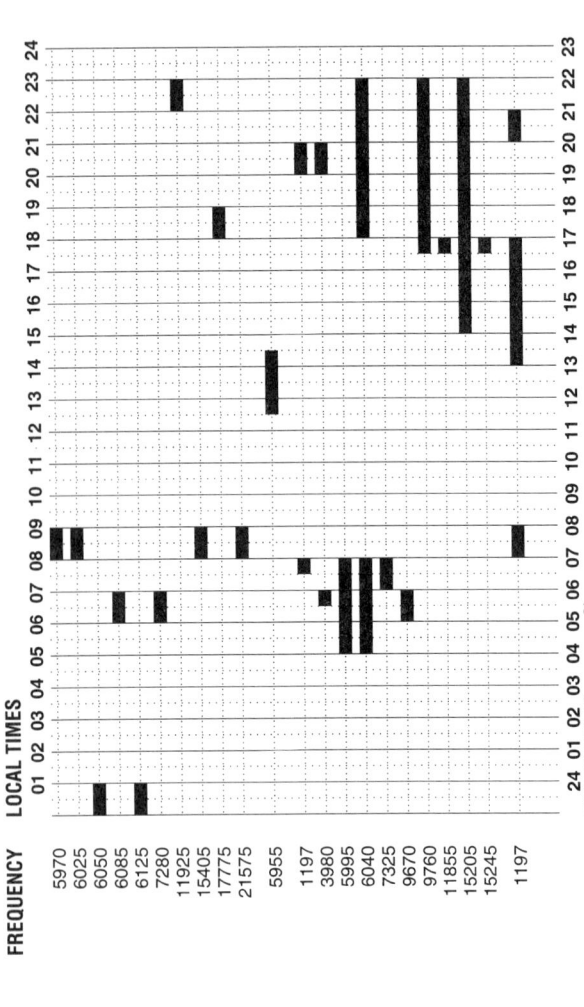

FREQUENCY		
5970 6025 6050 6085 6125 7280 11925 15405 17775 21575		Radio Japan
5955		Radio Netherlands
1197 3980 5995 6040 7325 9670 9760 11855 15205 15245		Voice of America
1197		Voice of America - VOA Europe

LOCAL TIMES
UTC TIMES (Summer +2)

NEW YORK 141

NEW YORK, USA

Hours difference from UTC (GMT): -5 (Summer -4)
Main business languages: English, Spanish
Electricity: 60Hz, 110V
Currency: U.S. Dollar
Telephone country prefix and area code: 1 (212)

LOCAL AM & FM STATIONS	FREQUENCY
WFAN Sports.	660
WOR Newstalk.	710
WNYC News, Talk (non-commercial station).	820
WCBS All news.	880
WINS All news.	1010
WQXR Fine arts, classical music.	1560
WNYC-FM News, fine arts (non-commercial station), classical music.	FM 93.9

142 NEW YORK

WQXR-FM
Fine arts, classical music.

FREQUENCY
FM 96.3

INTERNATIONAL BROADCASTERS

BBC World Service

Frequency	
6175	
9515	
9590	
9915	
15220	
17840	

Christian Science Monitor World Service

Frequency	
7395	
9465	
9985	
13770	
15665	

Deutsche Welle

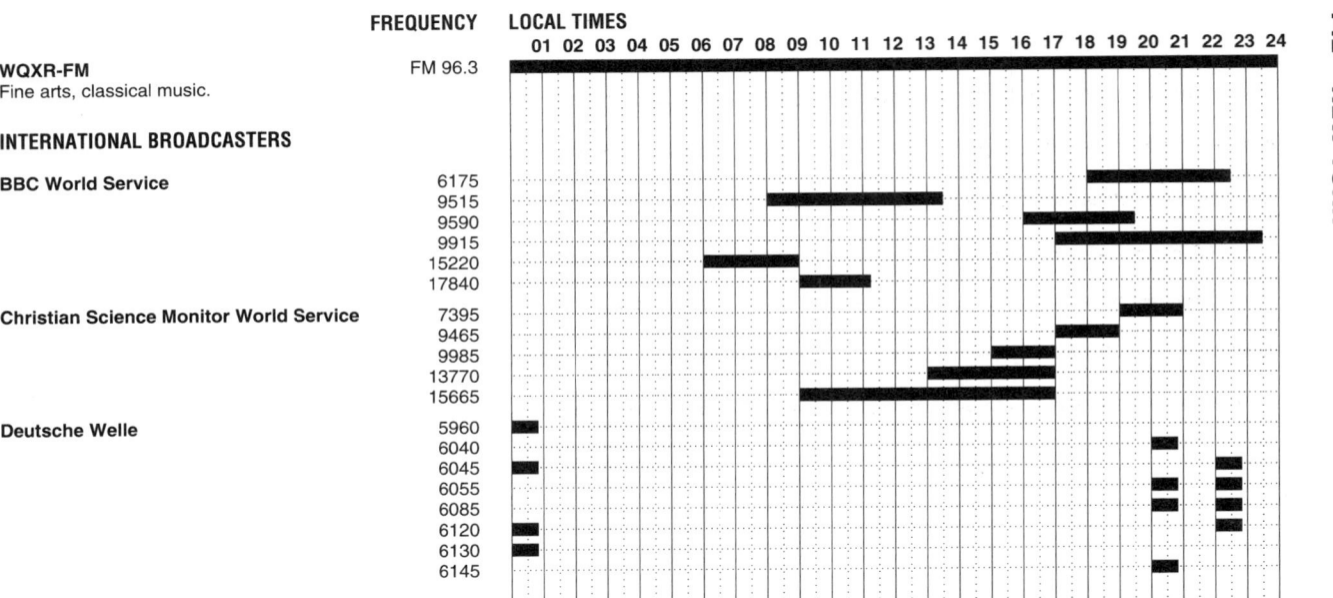

Frequency	
5960	
6040	
6045	
6055	
6085	
6120	
6130	
6145	

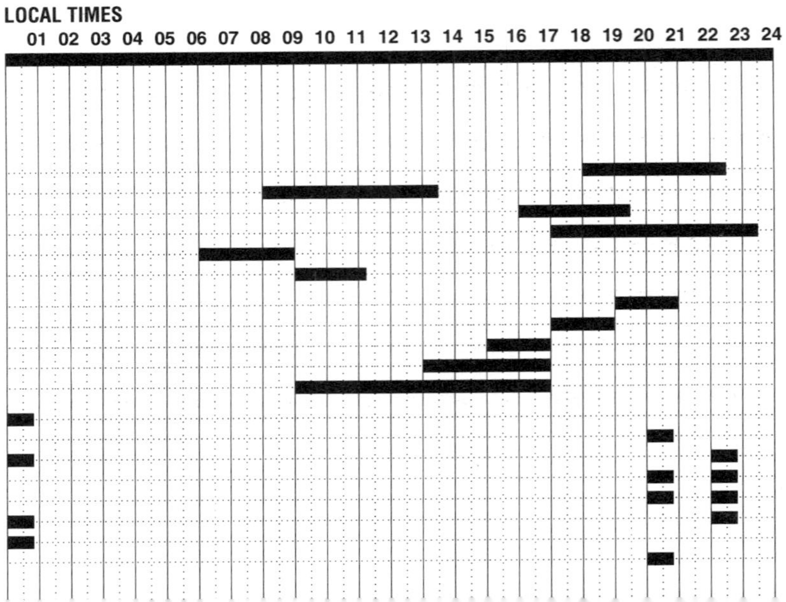

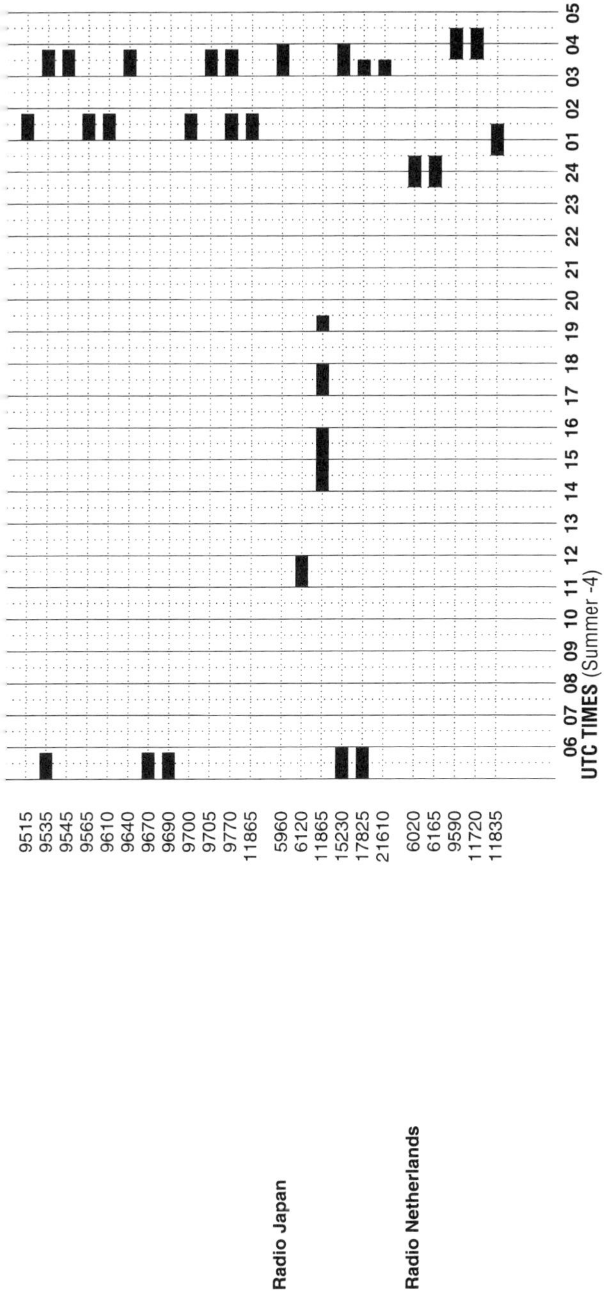

NEW YORK 143

Radio Japan

9515
9535
9545
9565
9610
9640
9670
9690
9700
9705
9770
11865

5960
6120
11865
15230
17825
21610

Radio Netherlands

6020
6165
9590
11720
11835

UTC TIMES (Summer -4)

OSLO, NORWAY

Hours difference from UTC (GMT): +1 (Summer +2)
Main business languages: Norwegian, English
Electricity: 50Hz, 220V
Currency: Norwegian Krone
Telephone country prefix and area code: 47 (2)

LOCAL AM & FM STATIONS	FREQUENCY
Radio Norway International Relay of program for listeners abroad.	FM 93.0
Voice of America - VOA Europe Program for young Europeans. News on the hour. Available on hotel cable system only.	FM 98.3
AFRTS U.S. Air Force station. News on the hour.	FM 105.5

INTERNATIONAL BROADCASTERS

BBC World Service

198
648
1296
6180
6195
7120
7325

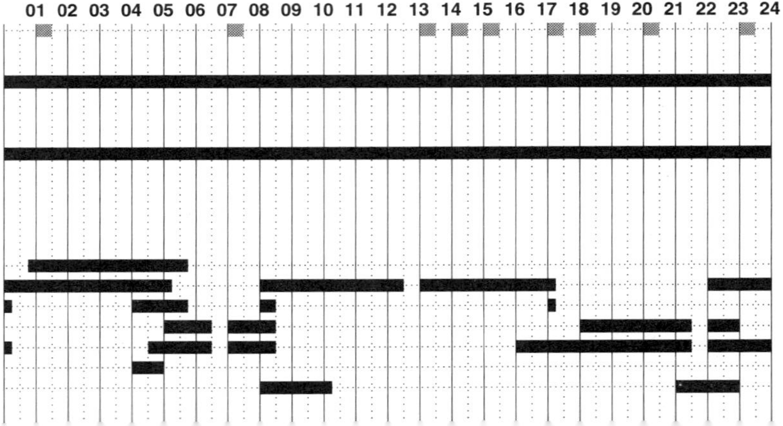

OSLO 145

Christian Science Monitor World Service

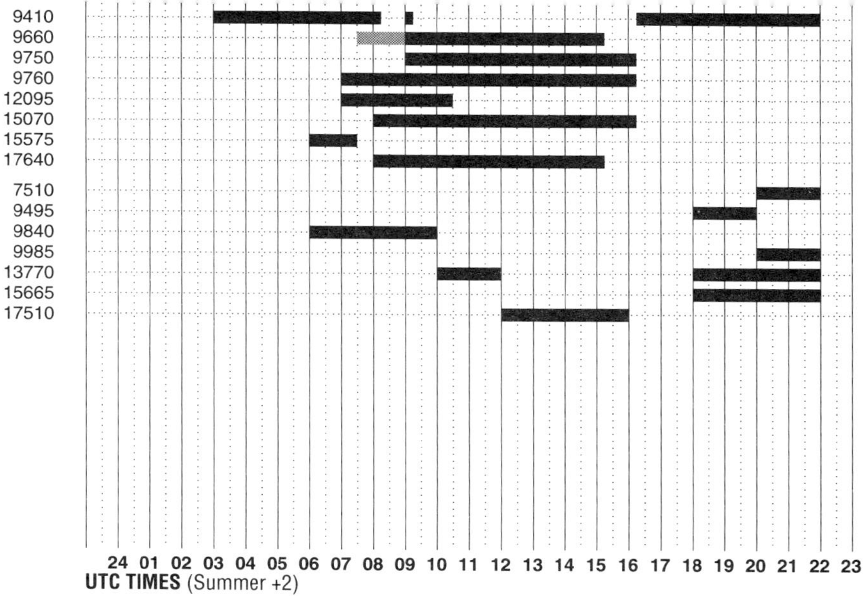

Frequency	UTC Times (Summer +2)
9410	
9660	
9750	
9760	
12095	
15070	
15575	
17640	
7510	
9495	
9840	
9985	
13770	
15665	
17510	

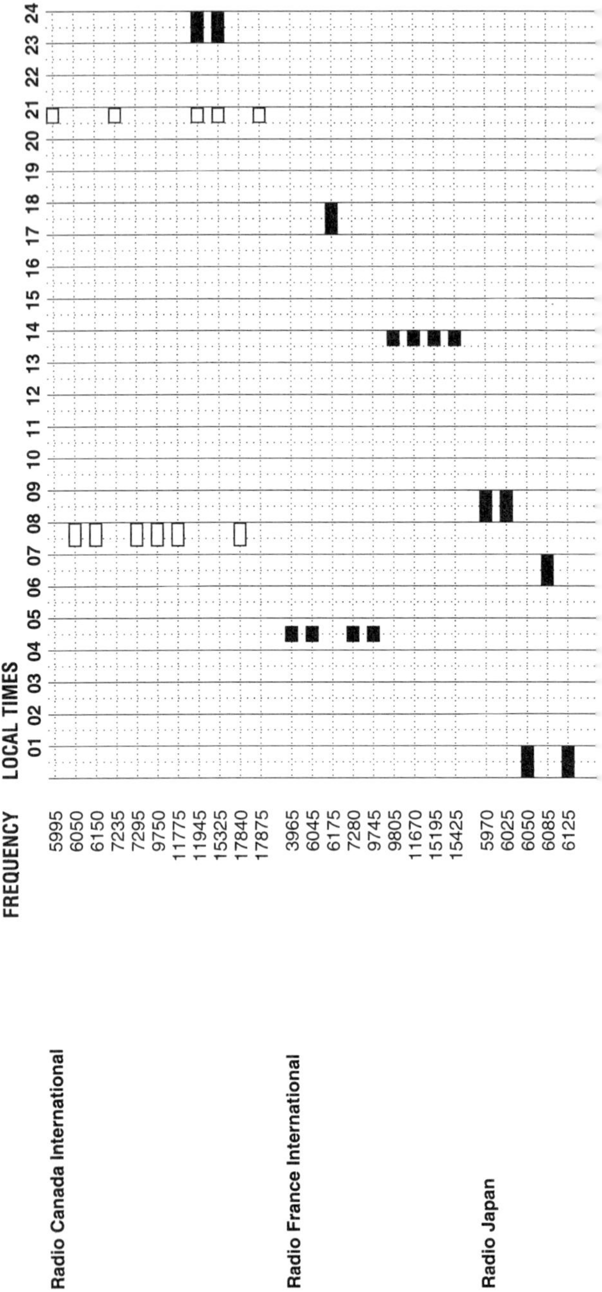

146 OSLO

Radio Canada International

Frequency	Local Times
5995	07–08
6050	07–08
6150	
7235	20–21
7295	07–08
9750	07–08
11775	07–08
11945	20–21, 23–24
15325	20–21, 23–24
17840	
17875	07–08, 20–21

Radio France International

Frequency	Local Times
3965	04–05
6045	04–05
6175	17–18
7280	04–05
9745	04–05
9805	13–14
11670	13–14
15195	13–14
15425	

Radio Japan

Frequency	Local Times
5970	08–09
6025	08–09
6050	06–07
6085	01–02
6125	01–02

OSLO 147

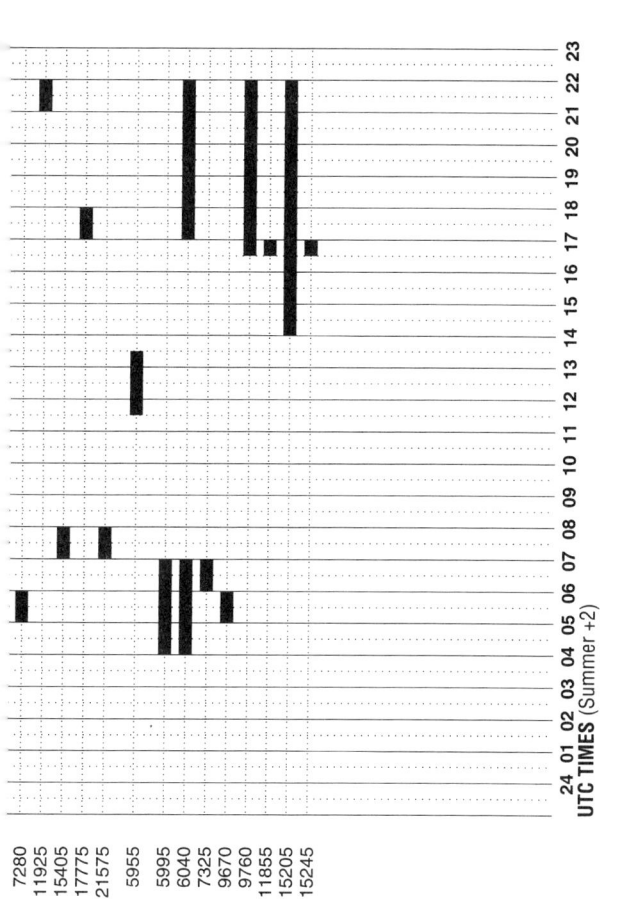

Radio Netherlands
- 7280
- 11925
- 15405
- 17775
- 21575

Voice of America
- 5955
- 5995
- 6040
- 7325
- 9670
- 9760
- 11855
- 15205
- 15245

UTC TIMES (Summer +2)

148 PARIS

PARIS, FRANCE
Hours difference from UTC (GMT): +1 (Summer +2)
Main business languages: French, English
Electricity: 50Hz, 220/240V
Currency: French Franc
Telephone country prefix and area code: 33 (1)

LOCAL AM & FM STATIONS

	FREQUENCY
BBC Radio 4 British station. Main news 0800, 0900, 1400, 1800, 1900, 2300, 2400.	198
Radio France International Relay of program for overseas listeners.	738

INTERNATIONAL BROADCASTERS

BBC World Service

	198
	648
	1296
	6180
	6195
	7120
	7325
	9410
	9660

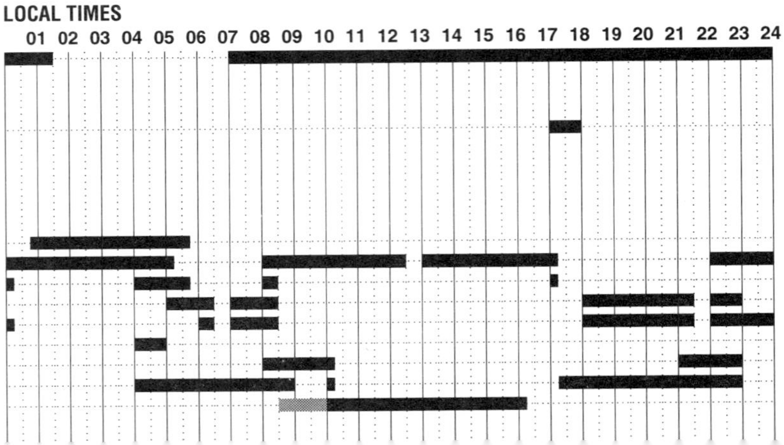

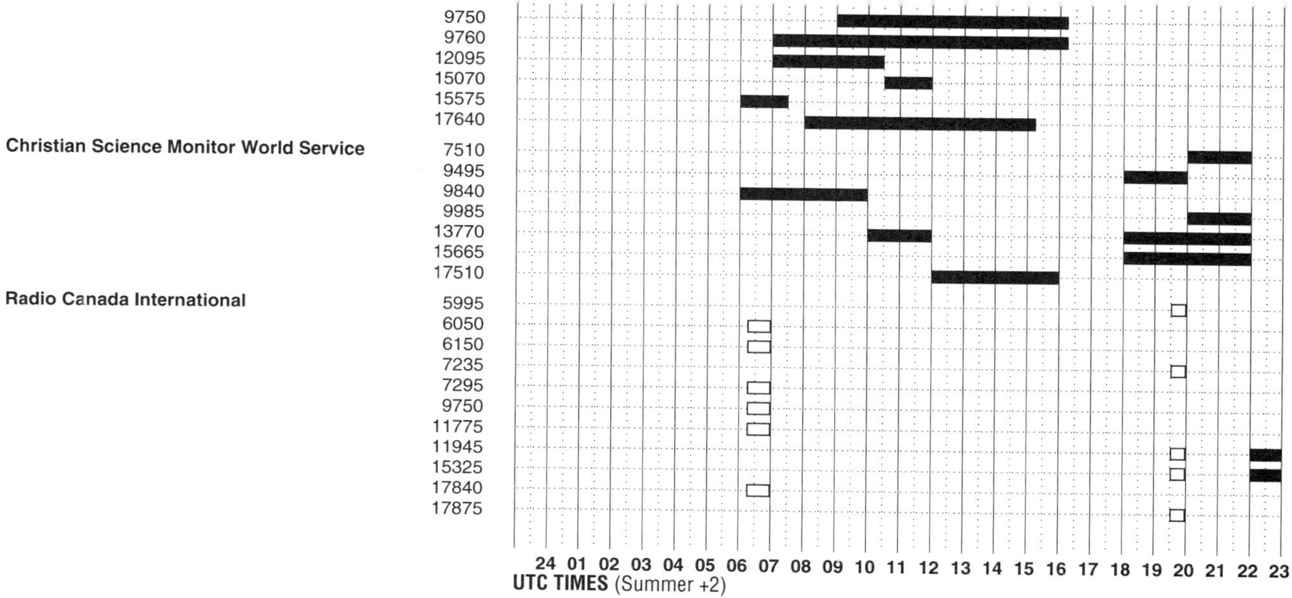

PARIS **149**

Christian Science Monitor World Service

Frequency
9750
9760
12095
15070
15575
17640
7510
9495
9840
9985
13770
15665
17510

Radio Canada International

Frequency
5995
6050
6150
7235
7295
9750
11775
11945
15325
17840
17875

UTC TIMES (Summer +2)

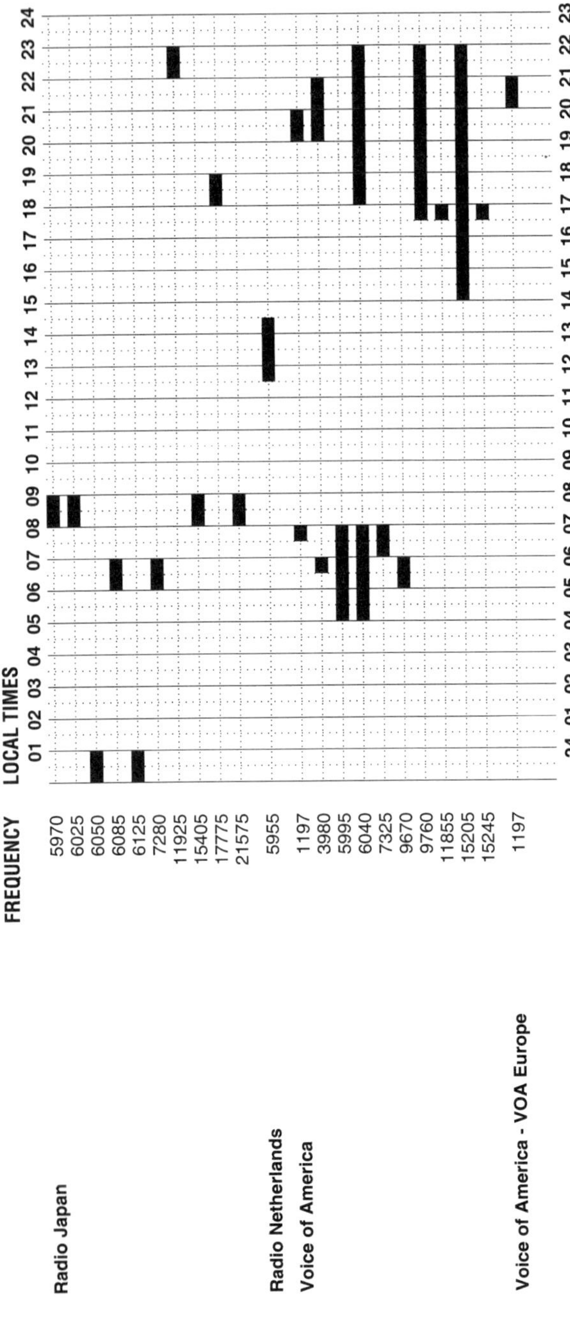

150 PARIS

Station	Frequency
Radio Japan	5970
	6025
	6050
	6085
	6125
	7280
	11925
	15405
	17775
	21575
Radio Netherlands	5955
Voice of America	1197
	3980
	5995
	6040
	7325
	9670
	9760
	11855
	15205
	15245
Voice of America - VOA Europe	1197

LOCAL TIMES / UTC TIMES (Summer +2)

PERTH 151

PERTH, AUSTRALIA
Hours difference from UTC (GMT): +8
Main business languages: English
Electricity: 50Hz, 240/415V
Currency: Australian Dollar
Telephone country prefix and area code: 61 (9)

LOCAL AM & FM STATIONS | FREQUENCY

ABC Metropolitan Service — 720
Regional non-commercial station.

ABC Radio National — 810
Non-commercial station.

6PR — 882
Commercial station. News on the hour.

6KY — 1206
Commercial station. News on the hour.

6UVS — FM 92.1
University station. Relays BBC News 0600, 0700, 0800.

6NOW — FM 96.1
Commercial station. News on the hour.

ABC FM Radio Service — FM 97.5
Non-commercial station.

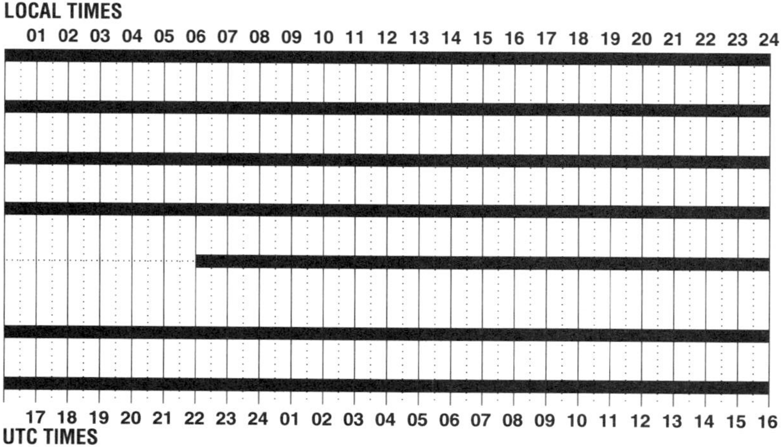

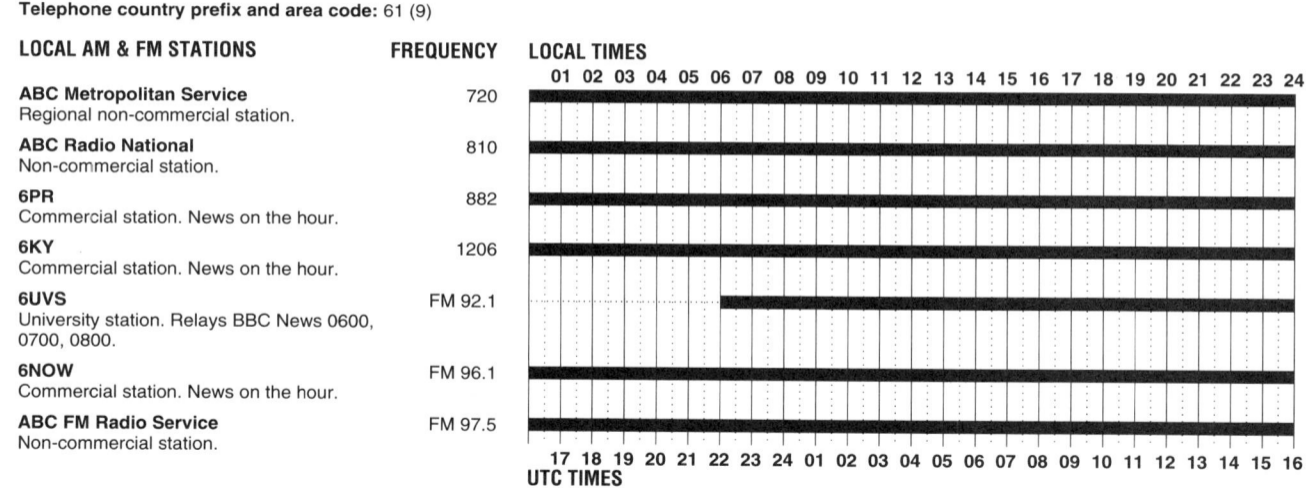

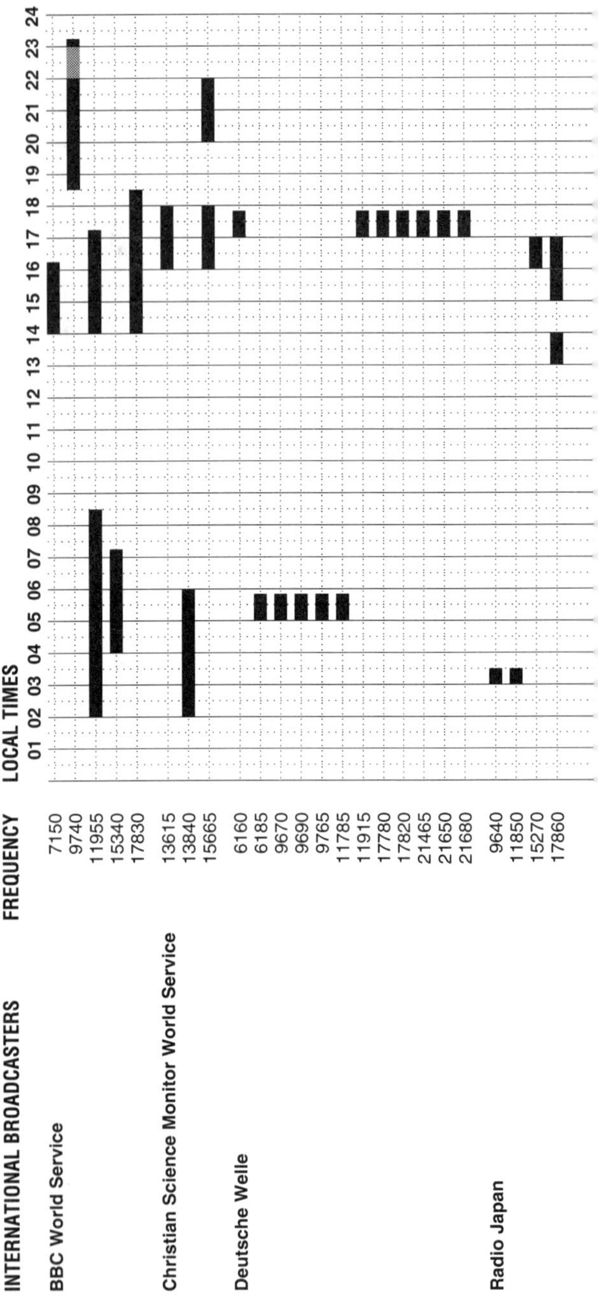

152 PERTH

INTERNATIONAL BROADCASTERS	FREQUENCY
BBC World Service	7150
	9740
	11955
	15340
	17830
Christian Science Monitor World Service	13615
	13840
	15665
Deutsche Welle	6160
	6185
	9670
	9690
	9765
	11785
	11915
	17780
	17820
	21465
	21650
	21680
Radio Japan	9640
	11850
	15270
	17860

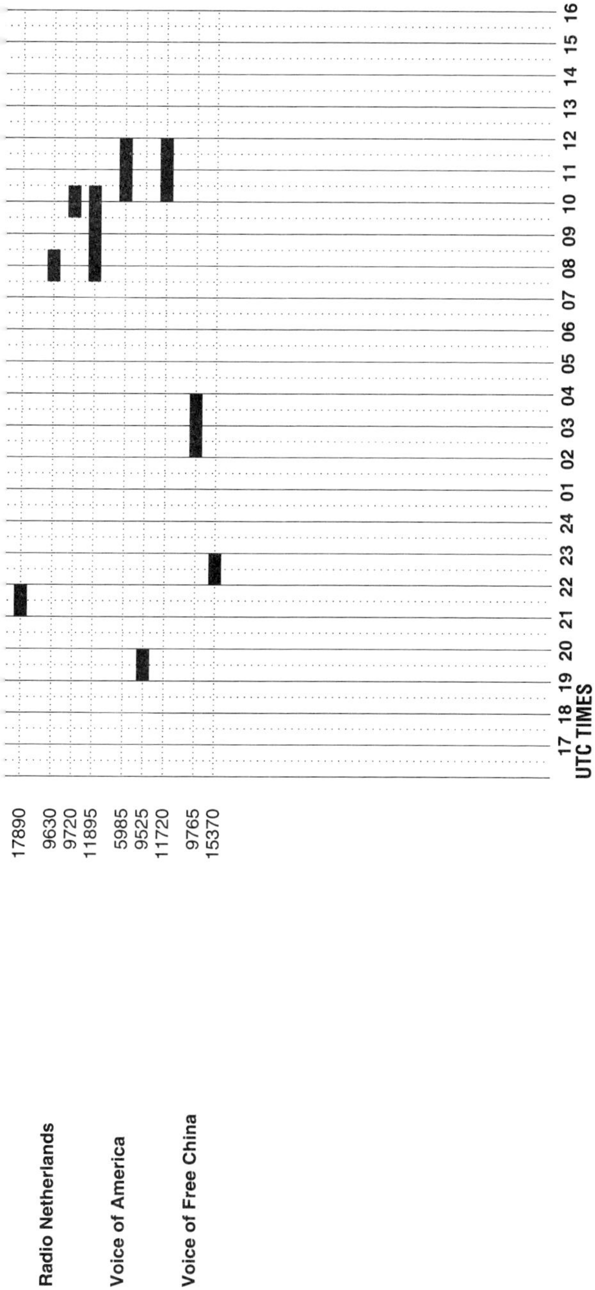

PERTH 153

Station	Frequency	UTC Times
Radio Netherlands	17890	21–22
	9630	08–09
	9720	09–10
	11895	08–10
Voice of America	5985	10–12
	9525	19–20
	11720	10–12
Voice of Free China	9765	02–04, 22–23
	15370	

RIO DE JANEIRO, BRAZIL

Hours difference from UTC (GMT): -3 (Summer -2)
Main business languages: Portuguese, English
Electricity: 60Hz, 110/220V
Currency: Cruzado
Telephone country prefix and area code: 55 (21)

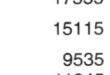

INTERNATIONAL BROADCASTERS	FREQUENCY
BBC World Service	6005
	9915
	11750
	15190
	15220
	15260
	17840
Christian Science Monitor World Service	9455
	9870
	13760
	17555
HCJB, Ecuador	15115
Radio Canada International	9535
	11845
	11940
	13720

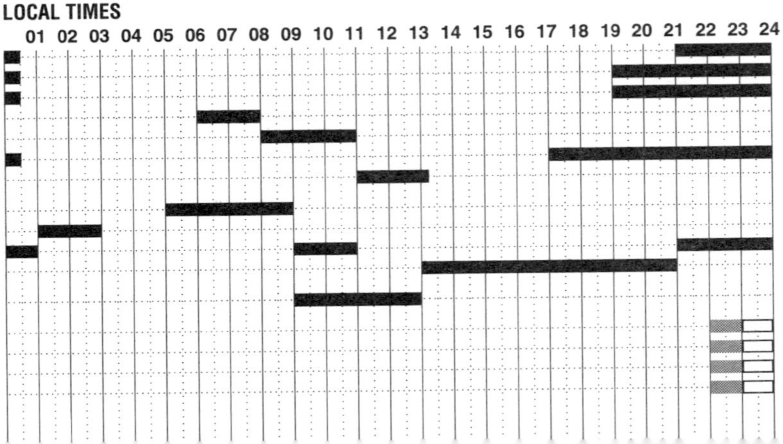

RIO DE JANEIRO 155

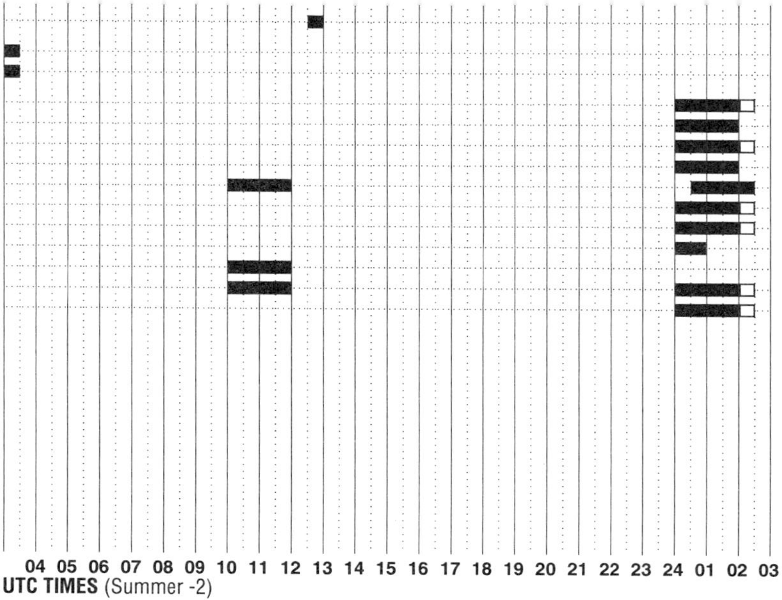

156　RIYADH

RIYADH, SAUDI ARABIA
Hours difference from UTC (GMT): +3
Main business languages: Arabic, English
Electricity: 50Hz, voltage varies
Currency: Riyal
Telephone country prefix and area code: 966 (1)

LOCAL AM & FM STATIONS	FREQUENCY
BSKSA English Service State-run station.	1422 FM 97.7

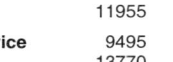

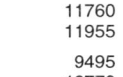

INTERNATIONAL BROADCASTERS	
BBC World Service	639
	702
	720
	1323
	1413
	5965
	6195
	9580
	11760
	11955
Christian Science Monitor World Service	9495
	13770

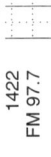

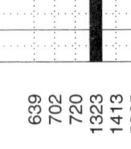

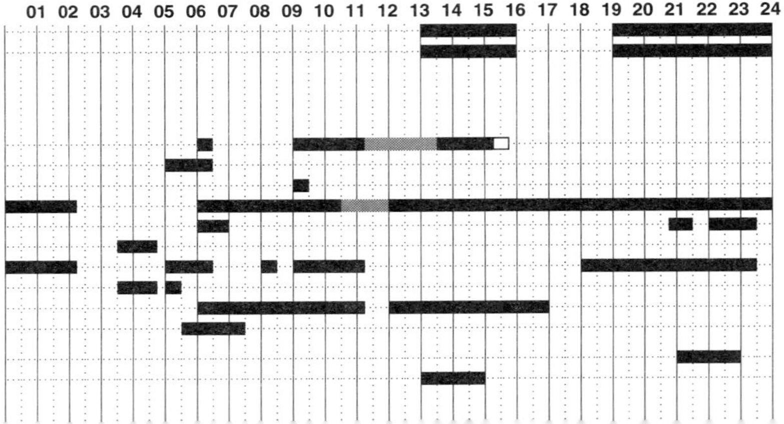

RIYADH 157

Radio Canada International	17510
	21640
	15275
Radio Japan	5970
	6025
	6050
	6085
	6125
	7280
	11925
	15405
	17775
	21575

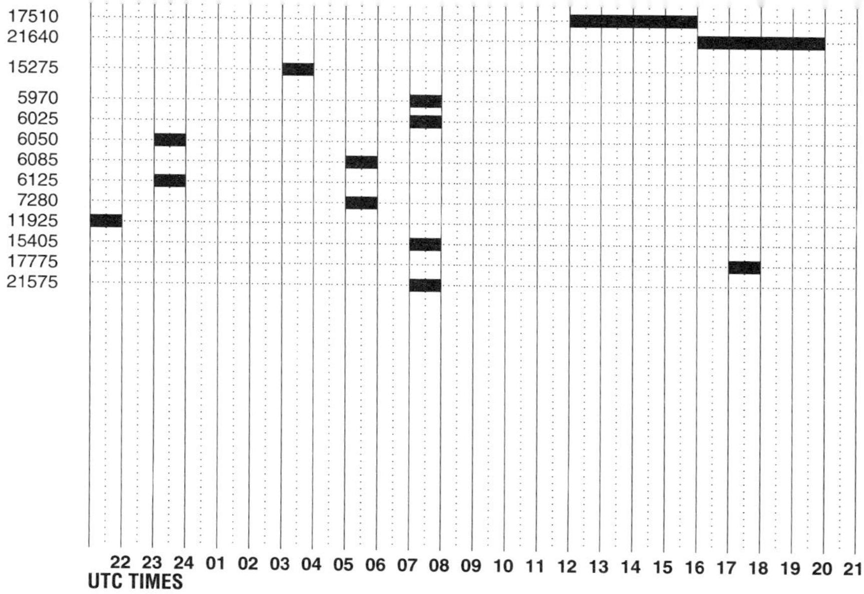

158 RIYADH

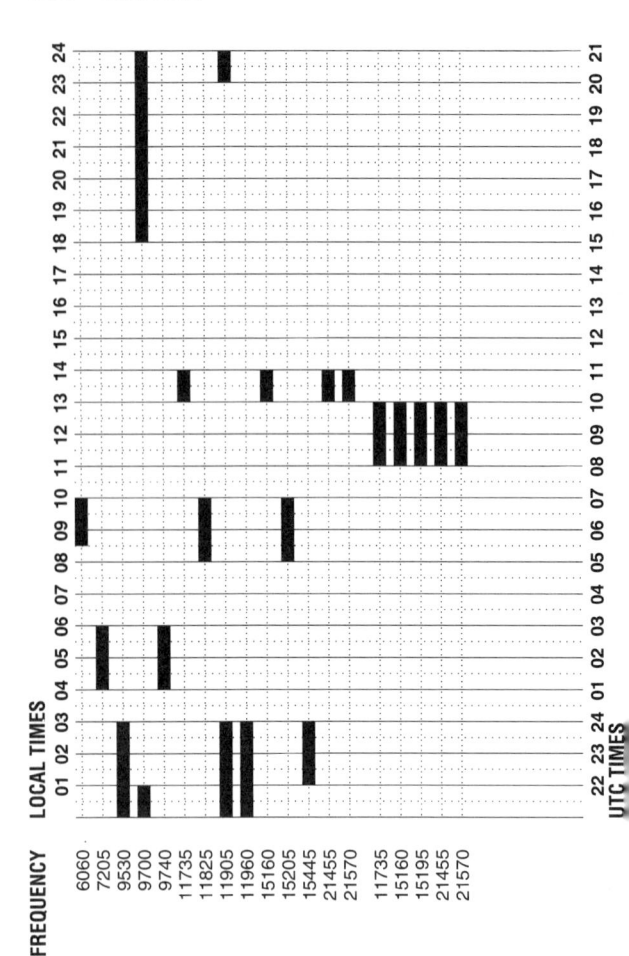

Voice of America

FREQUENCY	LOCAL TIMES
6060	08–09
7205	17–24
9530	04–05
9700	02
9740	02–03
11735	13–14
11825	01–02
11905	23–01
11960	23–01
15160	06–07, 13–14
15205	13–14
15445	00–01
21455	06–07
21570	

Voice of America - VOA Europe

FREQUENCY	LOCAL TIMES
11735	08–12
15160	08–12
15195	08–12
21455	08–12
21570	08–12

ROME, ITALY

Hours difference from UTC (GMT): +1 (Summer +2)
Main business languages: Italian, English
Electricity: 50Hz, voltage varies
Currency: Italian Lira
Telephone country prefix and area code: 39 (6)

LOCAL AM & FM STATIONS	FREQUENCY
Vatican Radio Programs in English and other languages.	526
RAI Night Program News summary at 3 minutes past each hour.	846
Vatican Radio Programs in English and other languages.	FM 93.3 FM 105.0

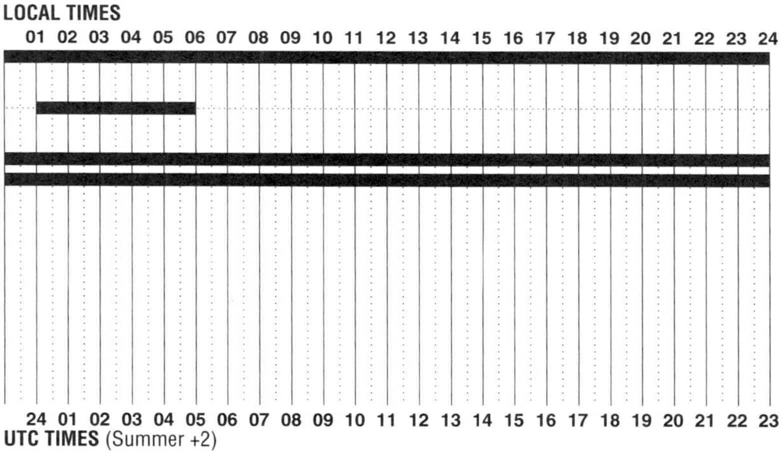

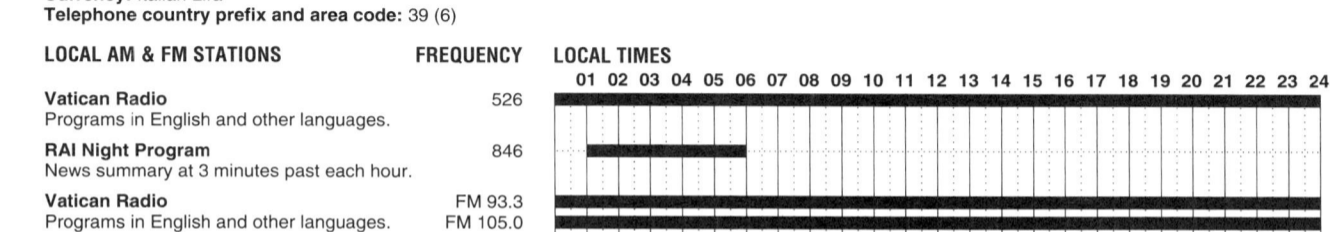

ROME 159

160 ROME

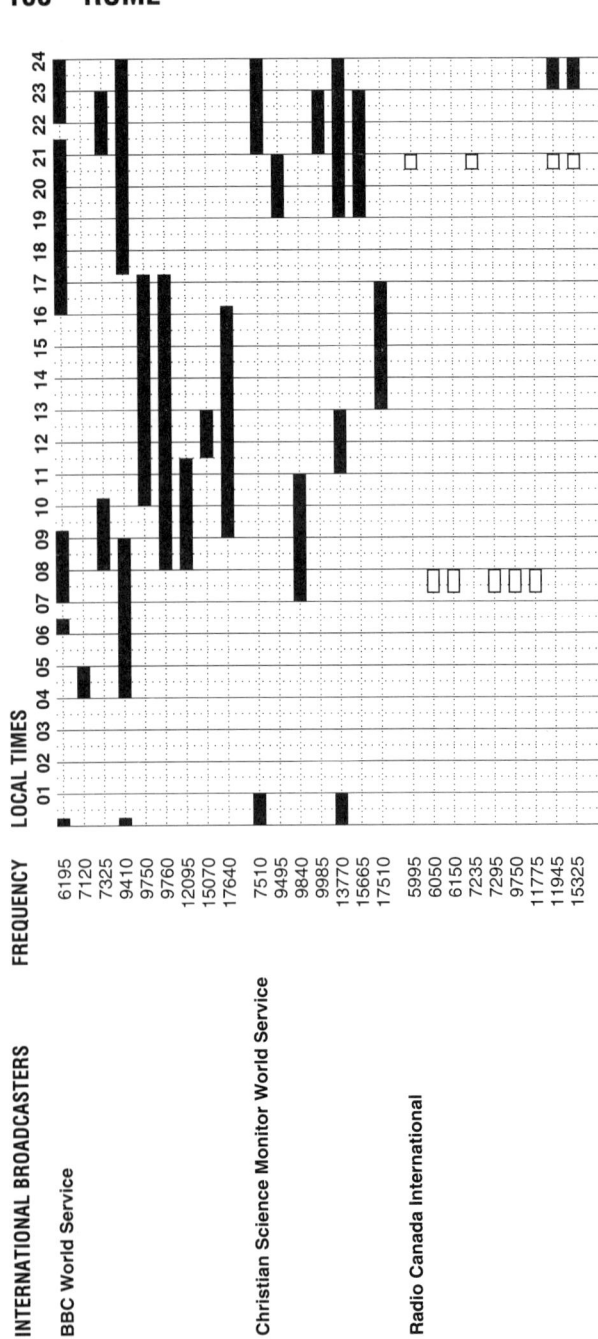

INTERNATIONAL BROADCASTERS	FREQUENCY
BBC World Service	6195
	7120
	7325
	9410
	9750
	9760
	12095
	15070
	17640
Christian Science Monitor World Service	7510
	9495
	9840
	9985
	13770
	15665
	17510
Radio Canada International	5995
	6050
	6150
	7235
	7295
	9750
	11775
	11945
	15325

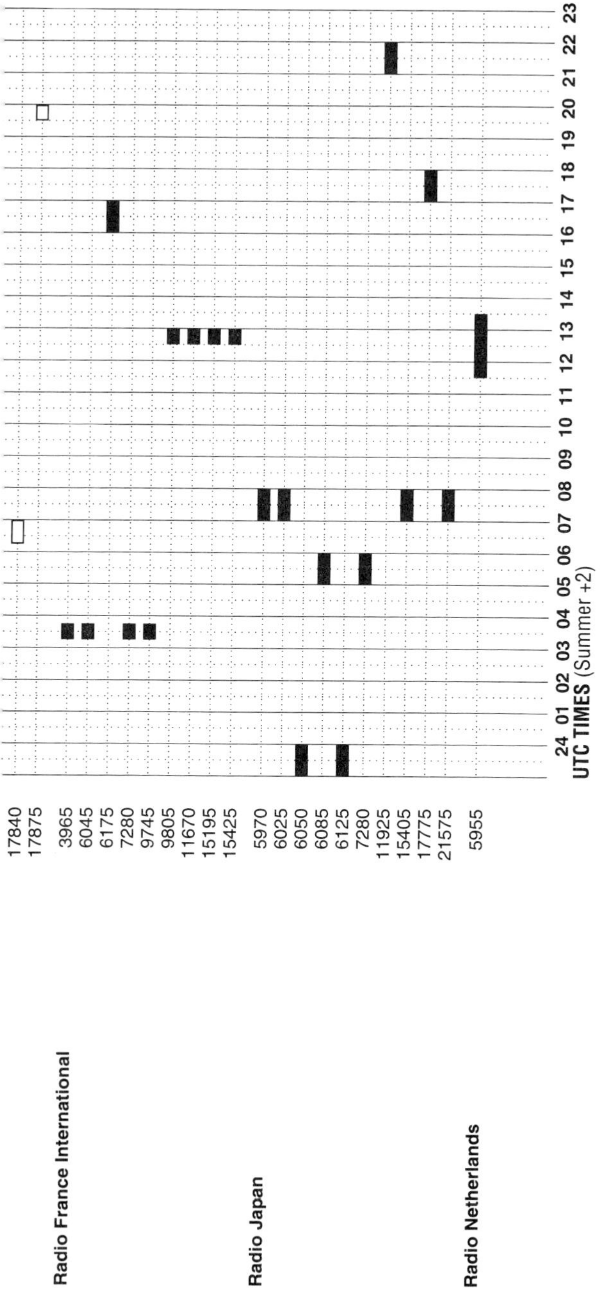

ROME 161

Radio France International

17840	
17875	
3965	
6045	
6175	
7280	
9745	
9805	
11670	
15195	
15425	

Radio Japan

5970	
6025	
6050	
6085	
6125	
7280	
11925	
15405	
17775	
21575	

Radio Netherlands

5955	

UTC TIMES (Summer +2)

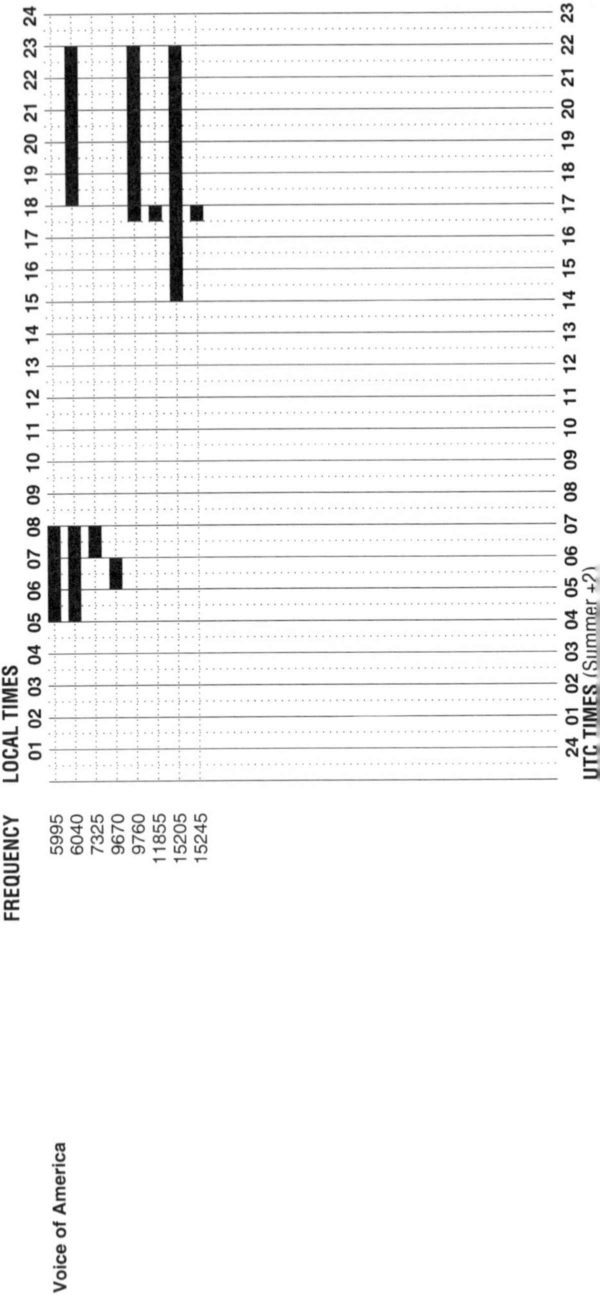

162 ROME

Voice of America

FREQUENCY	LOCAL TIMES	UTC TIMES (Summer +2)
5995	05–07	
6040	05–07	
7325	06–07	
9670	06	
9760	18–23	
11855	18, 21–23	
15205	15–18, 21–23	
15245	18	

SEOUL 163

SEOUL, KOREA
Hours difference from UTC (GMT): +9 (Summer +10)
Main business languages: Korean, English
Electricity: 60Hz, 100/200V
Currency: Won
Telephone country prefix and area code: 82 (2)

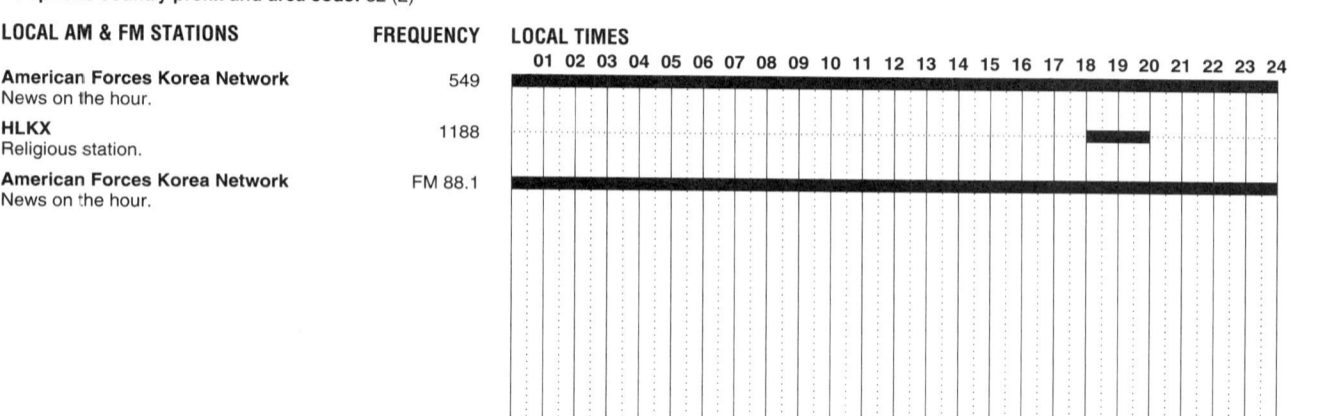

LOCAL AM & FM STATIONS	FREQUENCY
American Forces Korea Network News on the hour.	549
HLKX Religious station.	1188
American Forces Korea Network News on the hour.	FM 88.1

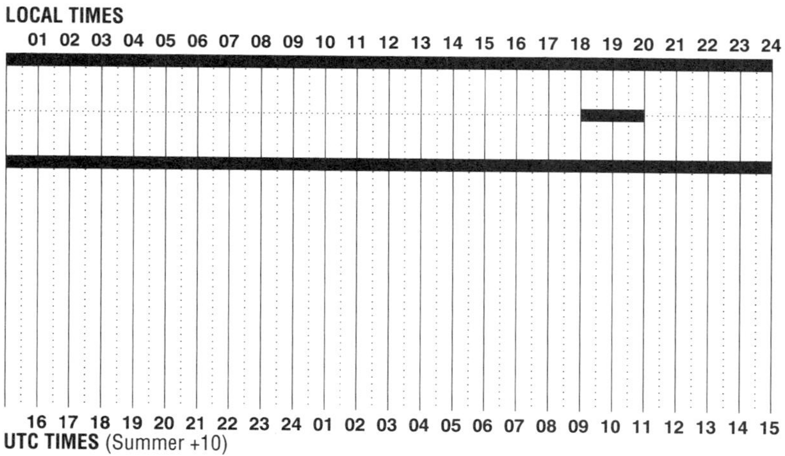

LOCAL TIMES
01 02 03 04 05 06 07 08 09 10 11 12 13 14 15 16 17 18 19 20 21 22 23 24

16 17 18 19 20 21 22 23 24 01 02 03 04 05 06 07 08 09 10 11 12 13 14 15
UTC TIMES (Summer +10)

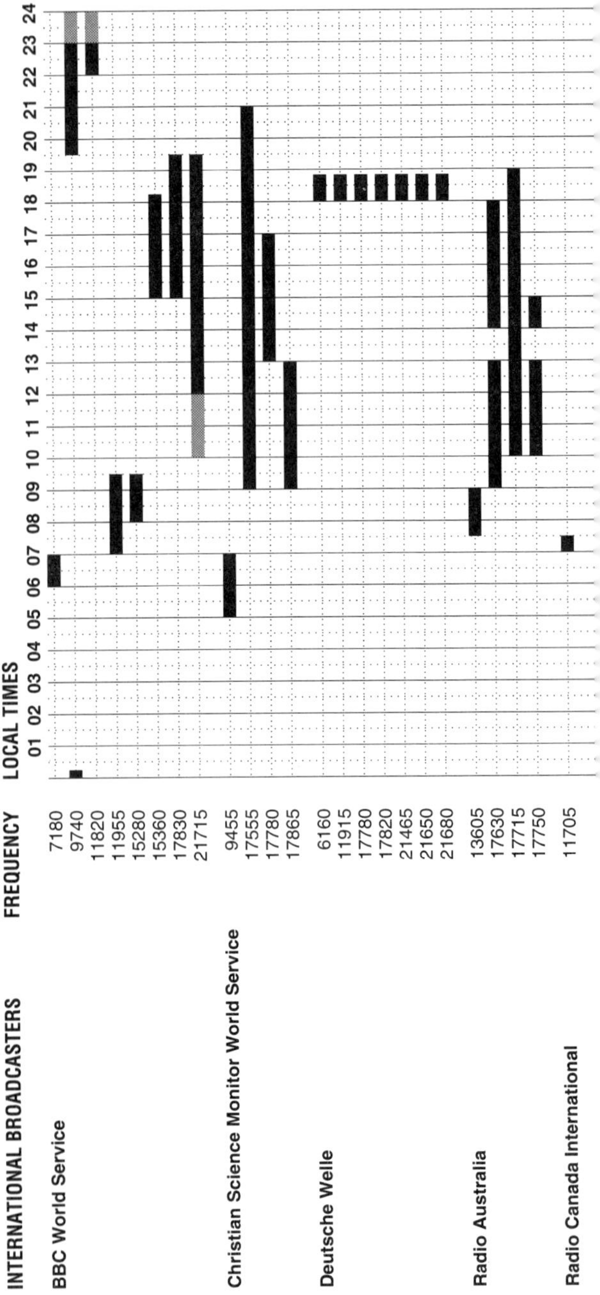

164 SEOUL

INTERNATIONAL BROADCASTERS	FREQUENCY
BBC World Service	7180
	9740
	11820
	11955
	15280
	15360
	17830
	21715
Christian Science Monitor World Service	9455
	17555
	17780
	17865
Deutsche Welle	6160
	11915
	17780
	17820
	21465
	21650
	21680
Radio Australia	13605
	17630
	17715
	17750
Radio Canada International	11705

LOCAL TIMES: 01 02 03 04 05 06 07 08 09 10 11 12 13 14 15 16 17 18 19 20 21 22 23 24

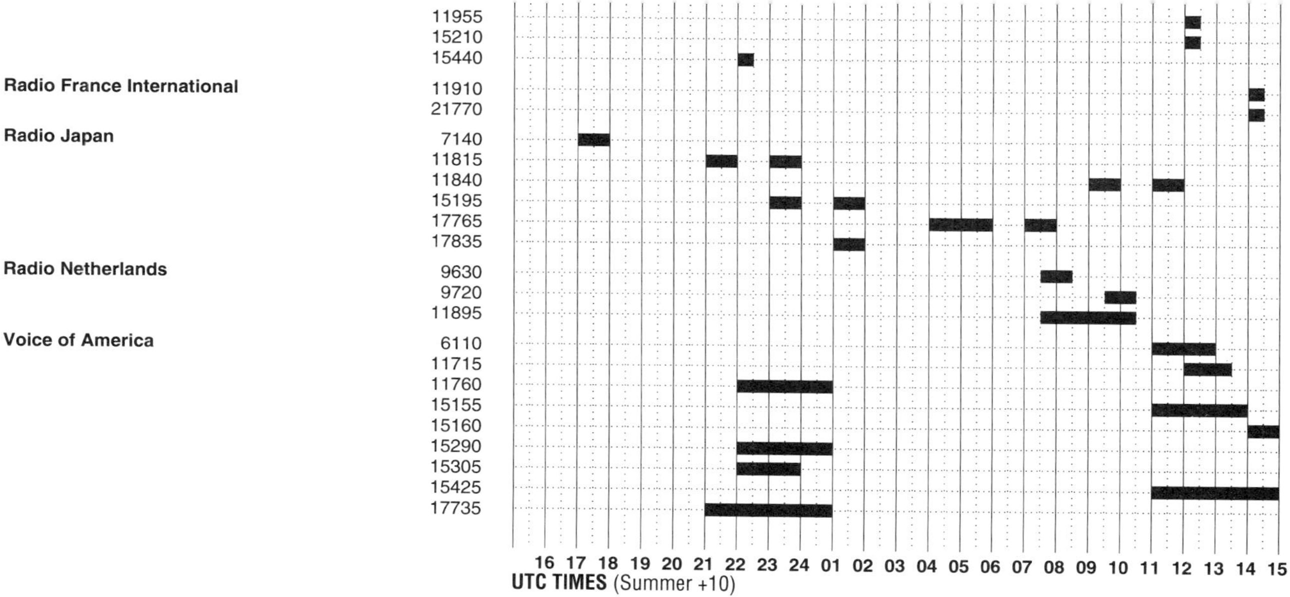

SEOUL 165

Station	Frequencies (kHz)
Radio France International	11955, 15210, 15440
Radio Japan	11910, 21770, 7140, 11815, 11840, 15195, 17765, 17835
Radio Netherlands	9630, 9720, 11895
Voice of America	6110, 11715, 11760, 15155, 15160, 15290, 15305, 15425, 17735

UTC TIMES (Summer +10)

SINGAPORE

Hours difference from UTC (GMT): +8
Main business languages: English
Electricity: 50Hz, 230V
Currency: Singapore Dollar
Telephone country prefix and area code: 65 ()

LOCAL AM & FM STATIONS

Station	Frequency
SBC Radio One — News on the hour.	630
BBC World Service — Satellite audio feed from London.	FM 88.9
SBC Radio One — News on the hour.	FM 90.5
SBC Radio Five — News 0600, 0700, 0800, 1200, 1300, 1700, 1800, 1900, 2200.	FM 92.4
Class 95FM — News at 0755, 0855, 1355, 1755, 1855, 2255.	FM 95.0
Perfect Ten — Non-commercial music station. News 0000, 0600, 1200, 1800.	FM 98.7

SINGAPORE

INTERNATIONAL BROADCASTERS

BBC World Service

Christian Science Monitor World Service

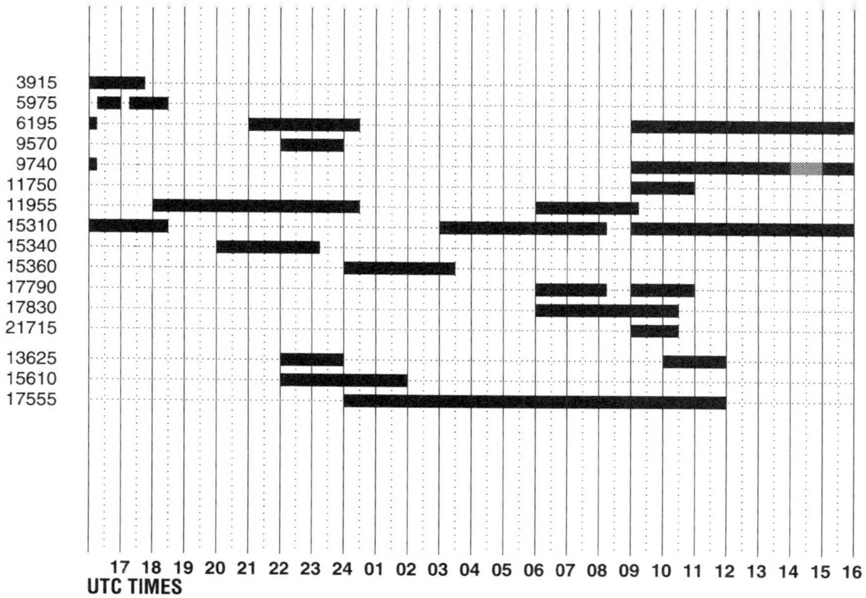

168 SINGAPORE

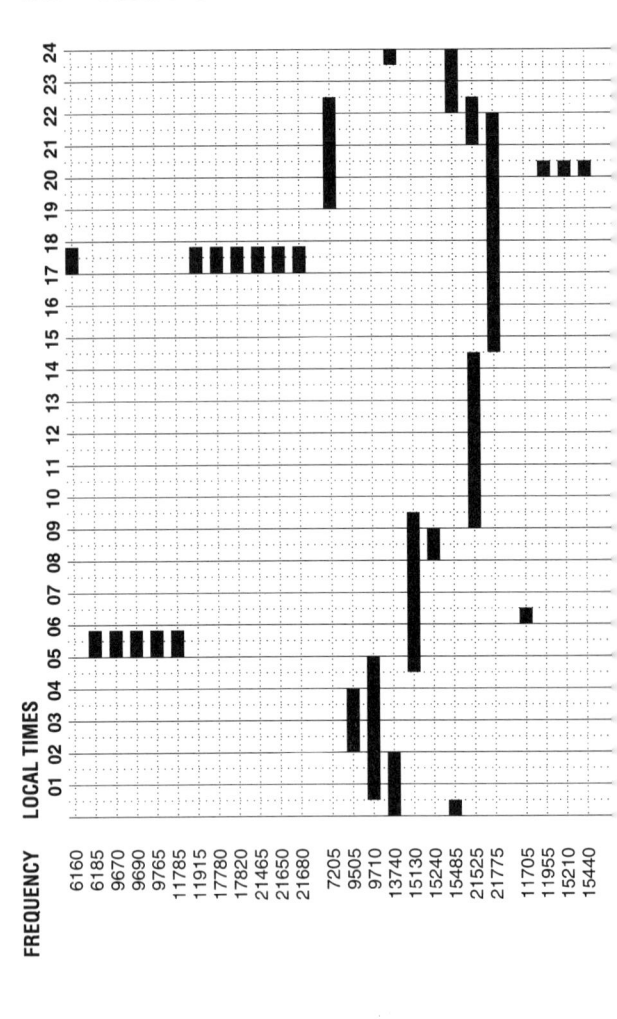

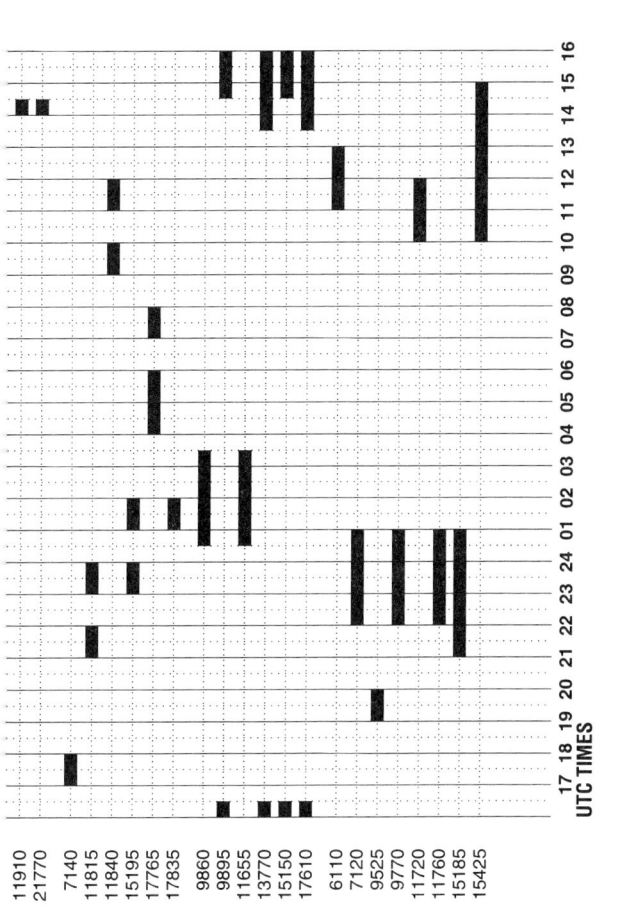

SINGAPORE 169

Station	Frequencies (kHz)
Radio France International	11910, 21770
Radio Japan	7140, 11815, 11840, 15195, 17765, 17835
Radio Netherlands	9860, 9895, 11655, 13770, 15150, 17610
Voice of America	6110, 7120, 9525, 9770, 11720, 11760, 15185, 15425

170 STOCKHOLM

STOCKHOLM, SWEDEN
Hours difference from UTC (GMT): +1 (Summer +2)
Main business languages: Swedish, English
Electricity: 50Hz, 220V
Currency: Swedish Krone
Telephone country prefix and area code: 46 (8)

LOCAL AM & FM STATIONS

	FREQUENCY
Radio Sweden Nordic news at start of program.	1179 FM 89.6
VOA Europe Operated by the Voice of America. News on the hour. Available on hotel cable system only.	FM 102.1

INTERNATIONAL BROADCASTERS

BBC World Service	198 1296 6180 6195 7120 7325 9410 9660 9750

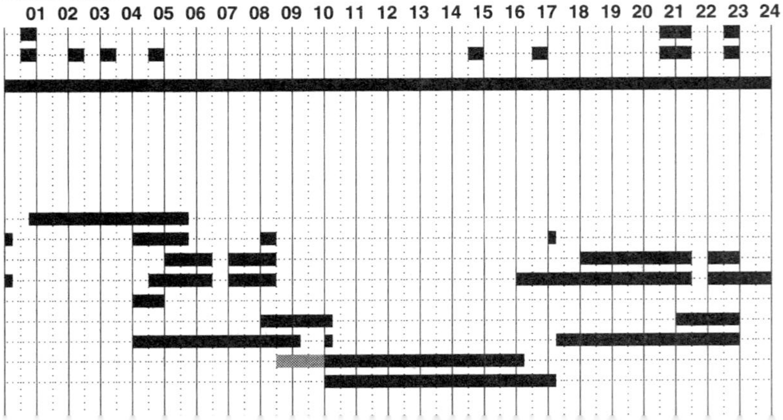

STOCKHOLM 171

Christian Science Monitor World Service

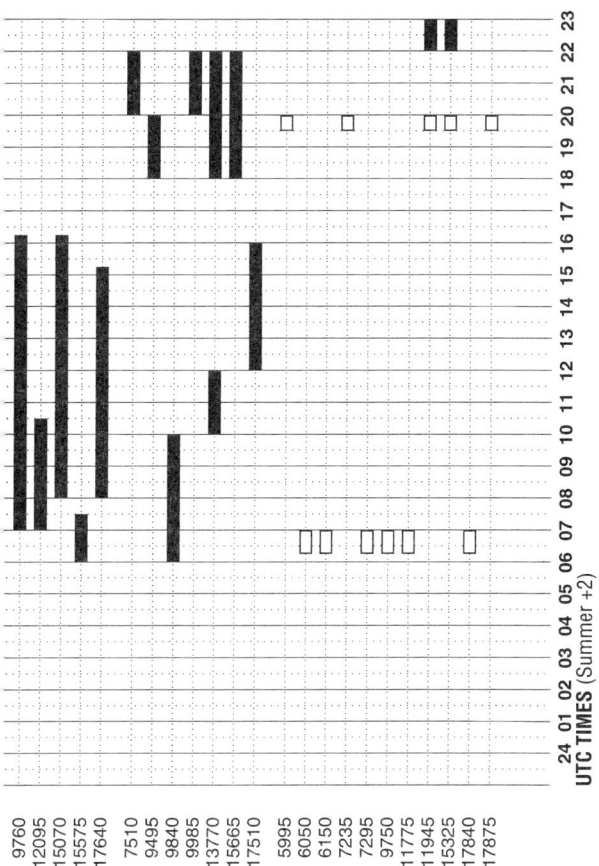

Frequency
9760
12095
15070
15575
17640
7510
9495
9840
9985
13770
15665
17510

Radio Canada International

Frequency
5995
6050
6150
7235
7295
9750
11775
11945
15325
17840
17875

UTC TIMES (Summer +2)

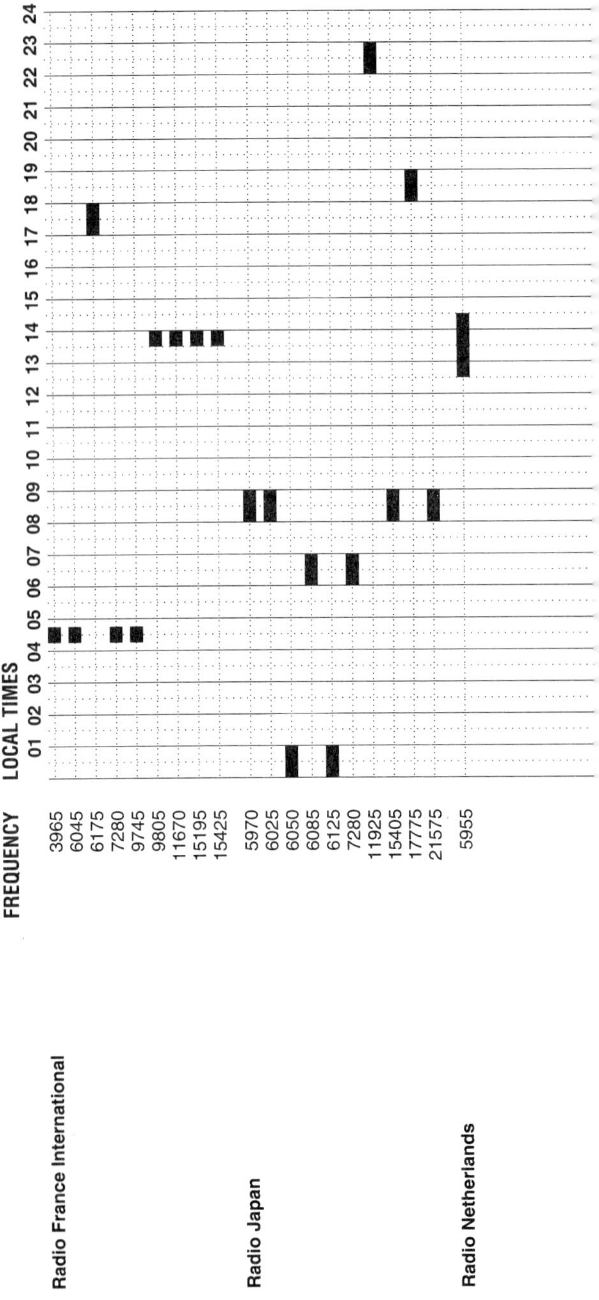

172 STOCKHOLM

FREQUENCY	
Radio France International	3965, 6045, 6175, 7280, 9745, 9805, 11670, 15195, 15425
Radio Japan	5970, 6025, 6050, 6085, 6125, 7280, 11925, 15405, 17775, 21575
Radio Netherlands	5955

STOCKHOLM 173

Frequency	Schedule
5995	
6040	
7325	
9670	
11855	
15205	
15245	

UTC TIMES (Summer +2)

Voice of America

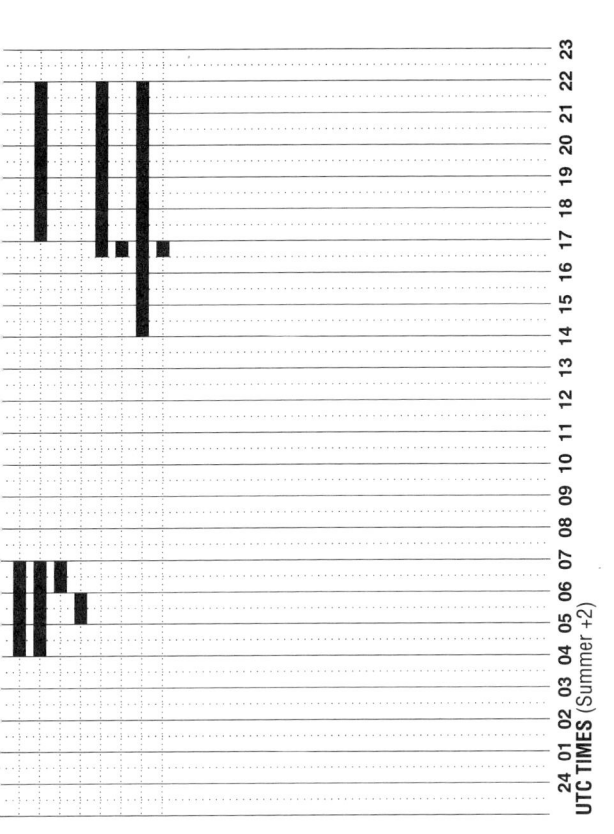

SYDNEY, AUSTRALIA

Hours difference from UTC (GMT): +10 (Summer +11)
Main business languages: English
Electricity: 50Hz, 240/415V
Currency: Australian Dollar
Telephone country prefix and area code: 61 (2)

LOCAL AM & FM STATIONS	FREQUENCY
ABC Radio National Non-commercial station.	576
ABC Metropolitan Service Regional non-commercial station.	702
2UE Commercial station. News on the hour.	954
2CH Commercial station. News on the hour.	1170
2SM Commercial station. News on the hour.	1269
ABC FM Radio Service Non-commercial station.	FM 92.9
2MMM-FM Commercial station. News on the hour.	FM 104.9

SYDNEY 175

INTERNATIONAL BROADCASTERS

BBC World Service

7150
9740
11955
15340
17830

Christian Science Monitor World Service

13615
15665

Deutsche Welle

6160
6185
9670
9690
9765
11785
11915
17780
17820
21465
21650
21680

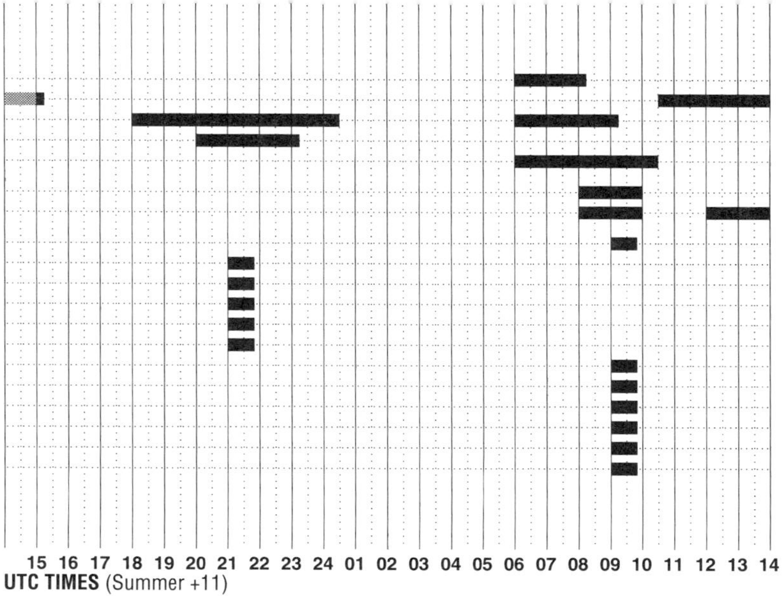

UTC TIMES (Summer +11)

176 SYDNEY

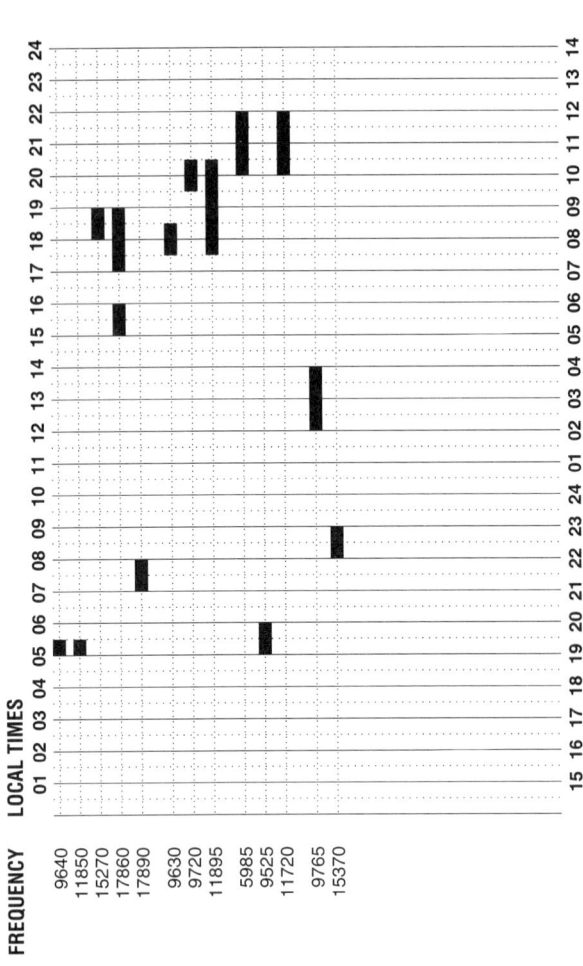

Station	Frequency	Local Time (approx.)
Radio Japan	9640	05–06
	11850	05–06
	15270	18–19
	17860	15–16
	17890	18–19
Radio Netherlands	9630	07–08
	9720	19–20
	11895	17–20
Voice of America	5985	20–21
	9525	19–21
	11720	20–22
Voice of Free China	9765	22–23
	15370	13–15

TAIPEI 177

TAIPEI, TAIWAN
Hours difference from UTC (GMT): +8
Main business languages: Chinese, English
Electricity: 60Hz, 110/220V
Currency: New Taiwan Dcllar
Telephone country prefix and area code: 886 (2)

LOCAL AM & FM STATIONS	FREQUENCY
International Community Radio News on the hour.	576 FM 100.7

INTERNATIONAL BROADCASTERS

BBC World Service	7180 9740 11820 11955 15280 15360 17830 21715

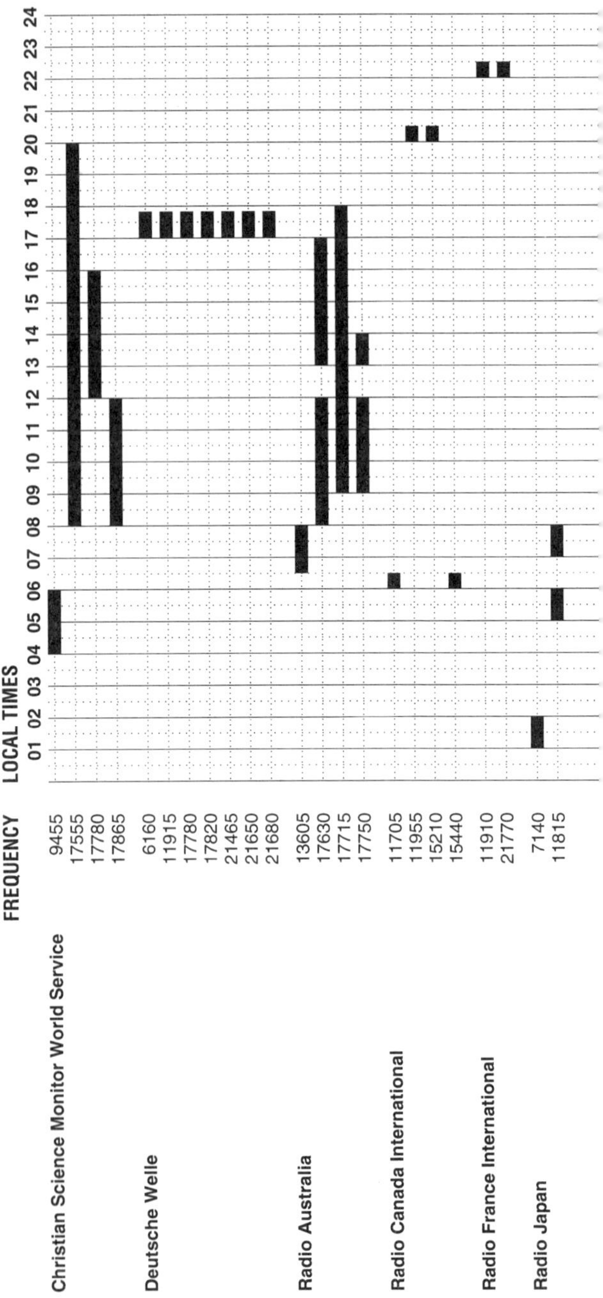

178 TAIPEI

Station	Frequency
Christian Science Monitor World Service	9455
	17555
	17780
	17865
Deutsche Welle	6160
	11915
	17780
	17820
	21465
	21650
	21680
Radio Australia	13605
	17630
	17715
	17750
Radio Canada International	11705
	11955
	15210
	15440
Radio France International	11910
	21770
Radio Japan	7140
	11815

TAIPEI 179

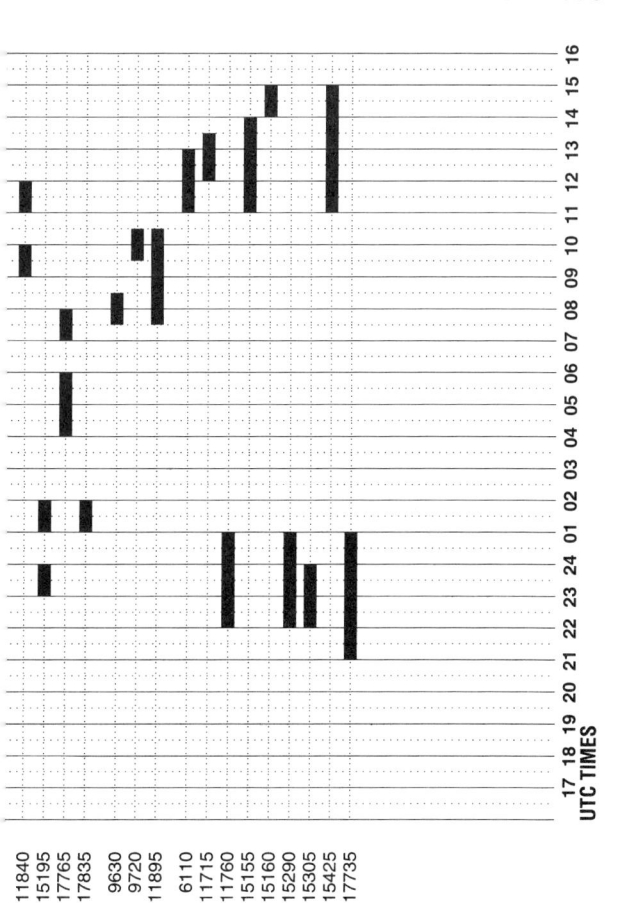

Radio Netherlands

Frequency
11840
15195
17765
17835

Voice of America

Frequency
9630
9720
11895
6110
11715
11760
15155
15160
15290
15305
15425
17735

UTC TIMES

180 TEL AVIV

TEL AVIV, ISRAEL
Hours difference from UTC (GMT): +2 (Summer +3)
Main business languages: Hebrew, English
Electricity: 50Hz, 230/400V
Currency: Shekel
Telephone country prefix and area code: 972 (3)

LOCAL AM & FM STATIONS | FREQUENCY

Israel Radio Program A — 576
News bulletin.

Voice of Peace — 1540
Commercial music station. Sometimes operates FM 100.0 reduced schedule.

INTERNATIONAL BROADCASTERS

BBC World Service
639
702
720
1323
1413
5965
6195
9580
11760

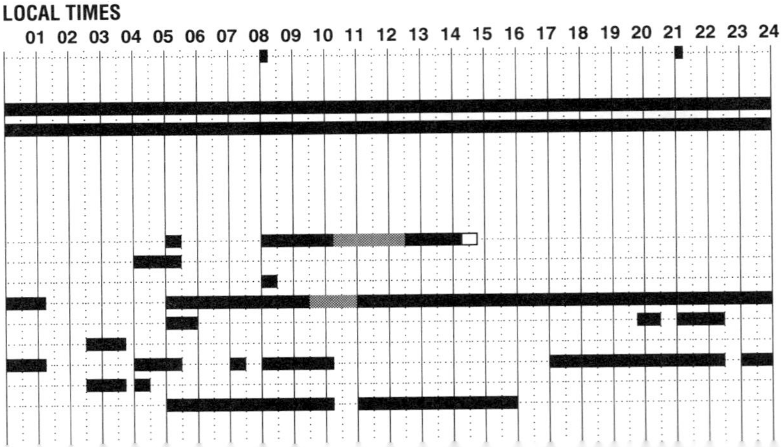

TEL AVIV

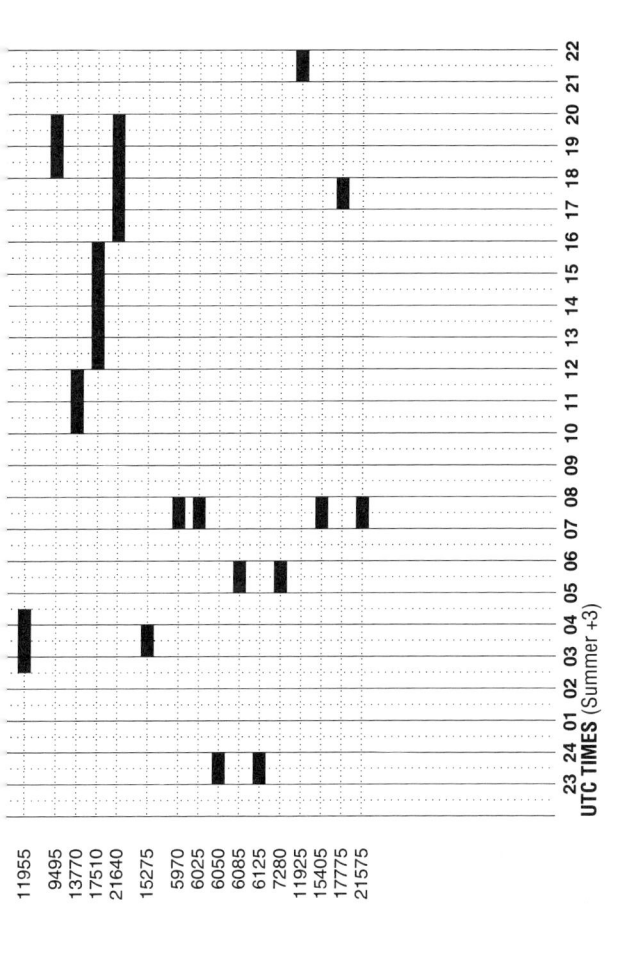

Christian Science Monitor World Service
- 11955
- 9495
- 13770
- 17510
- 21640

Radio Canada International
- 15275

Radio Japan
- 5970
- 6025
- 6050
- 6085
- 6125
- 7280
- 11925
- 15405
- 17775
- 21575

UTC TIMES (Summer +3)

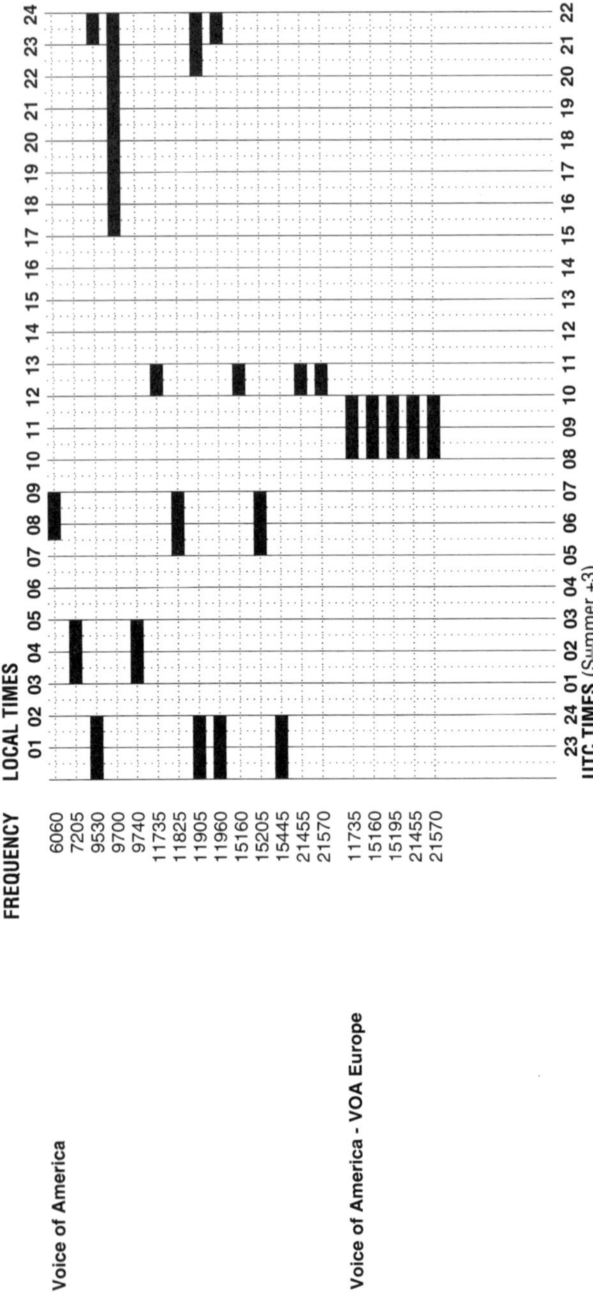

182 TEL AVIV

FREQUENCY		
Voice of America	6060 7205 9530 9700 9740 11735 11825 11905 11960 15160 15205 15445 21455 21570	
Voice of America - VOA Europe	11735 15160 15195 21455 21570	

LOCAL TIMES: 01 02 03 04 05 06 07 08 09 10 11 12 13 14 15 16 17 18 19 20 21 22 23 24

UTC TIMES (Summer +3): 23 24 01 02 03 04 05 06 07 08 09 10 11 12 13 14 15 16 17 18 19 20 21 22

TOKYO 183

TOKYO, JAPAN
Hours difference from UTC (GMT): +9
Main business languages: Japanese, English
Electricity: 60Hz, 110/220V
Currency: Yen
Telephone country prefix and area code: 81 (3)

LOCAL AM & FM STATIONS	FREQUENCY
Far East Network	810
U.S.Forces station. News on the hour.	

INTERNATIONAL BROADCASTERS

BBC World Service	7180
	9740
	11820
	11955
	15280
	15360
	17830
	21715
Christian Science Monitor World Service	9455
	17555
	17865

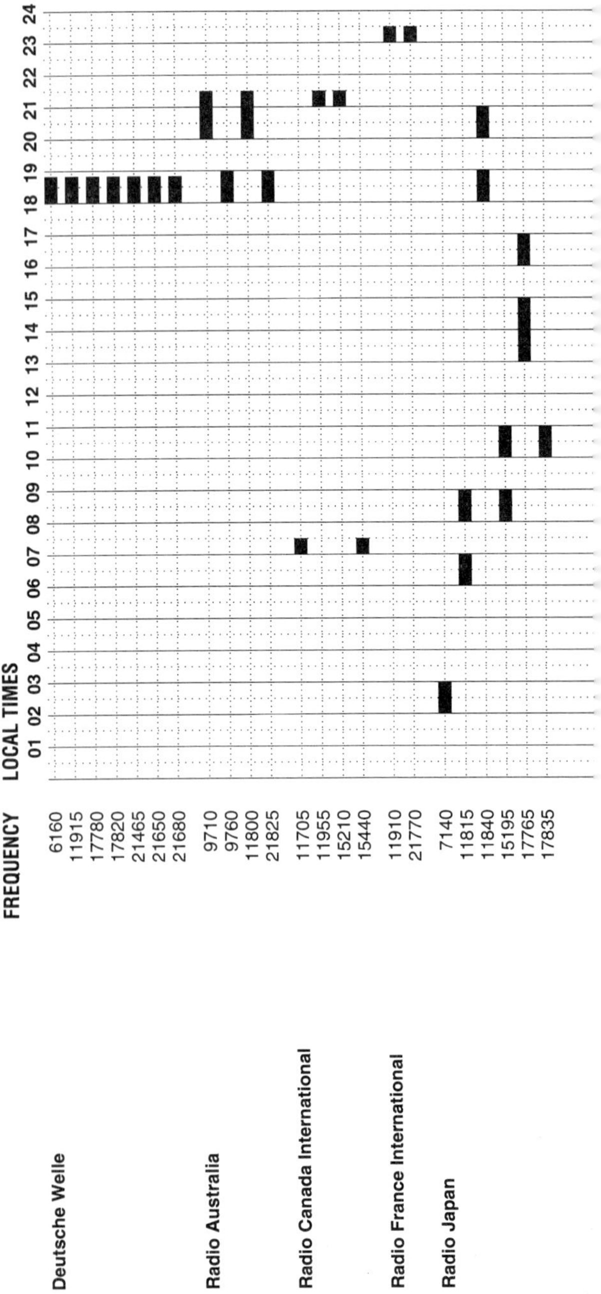

184 TOKYO

Broadcaster	Frequency	Local Times
Deutsche Welle	6160	18–19
	11915	18–19
	17780	18–19
	17820	18–19
	21465	18–19
	21650	18–19
	21680	18–19
Radio Australia	9710	19–20
	9760	20–21
	11800	19–20, 20–21
	21825	19–20
Radio Canada International	11705	07
	11955	21
	15210	21
	15440	07
Radio France International	11910	23
	21770	23
Radio Japan	7140	02–03
	11815	06, 18–19
	11840	20
	15195	08–09
	17765	14–15
	17835	10–11

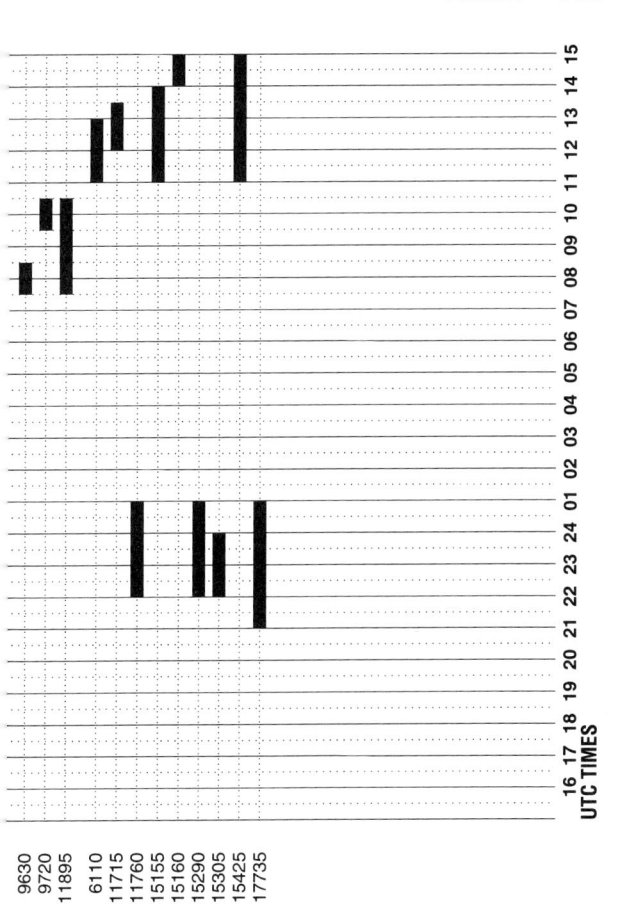

TOKYO 185

Radio Netherlands

9630
9720
11895

Voice of America

6110
11715
11760
15155
15160
15290
15305
15425
17735

UTC TIMES

186 TORONTO

TORONTO, CANADA
Hours difference from UTC (GMT): -5 (Summer -4)
Main business languages: English
Electricity: 60Hz, 115/230V
Currency: Canadian Dollar
Telephone country prefix and area code: 1 (416)

LOCAL AM & FM STATIONS	FREQUENCY	LOCAL TIMES 01–24
CKEY Commercial station. News on the hour.	590	
CFTR Commercial station.	680	
CBC English Radio Network Canadian Broadcasting Corporation. Non-commercial station.	740	
CFRB Commercial station. News on the hour.	1010	
CJCL Commercial station. News on the hour.	1430	
CHIN Commercial station. Programs in English and foreign languages.	1540	

TORONTO 187

CJRT-FM FM 91.1
Relay of BBC World Service News.

CBL-FM FM 94.1
Canadian Broadcasting Corporation.
Non-commercial station.

INTERNATIONAL BROADCASTERS

BBC World Service
6175
9515
9590
9915
15220
17840

Christian Science Monitor World Service
7395
9985
13770
15665

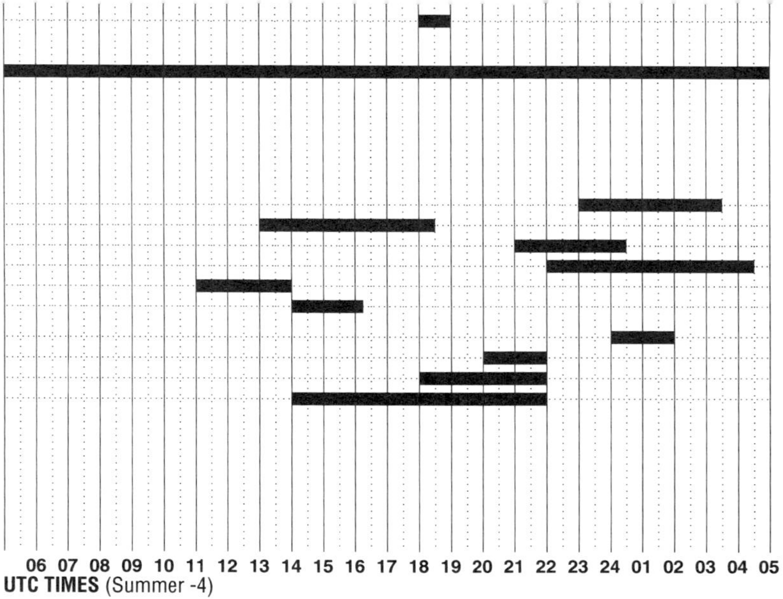

UTC TIMES (Summer -4)

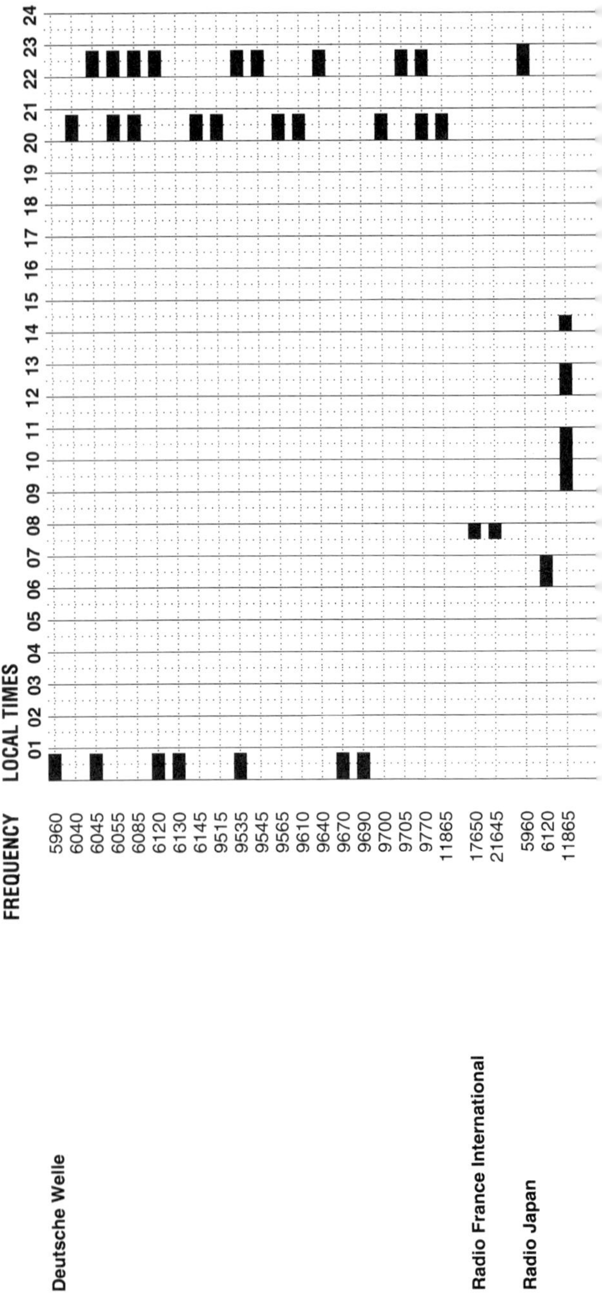

188 TORONTO

FREQUENCY	Deutsche Welle
5960	
6040	01, 20
6045	
6055	20
6085	20
6120	01, 22
6130	01, 22
6145	
9515	01
9535	20
9545	20
9565	22
9610	
9640	20
9670	01
9690	01
9700	
9705	20
9770	20
11865	22

Radio France International

| 17650 | 07 |
| 21645 | 07 |

Radio Japan

5960	06
6120	12, 22
11865	09, 12, 14

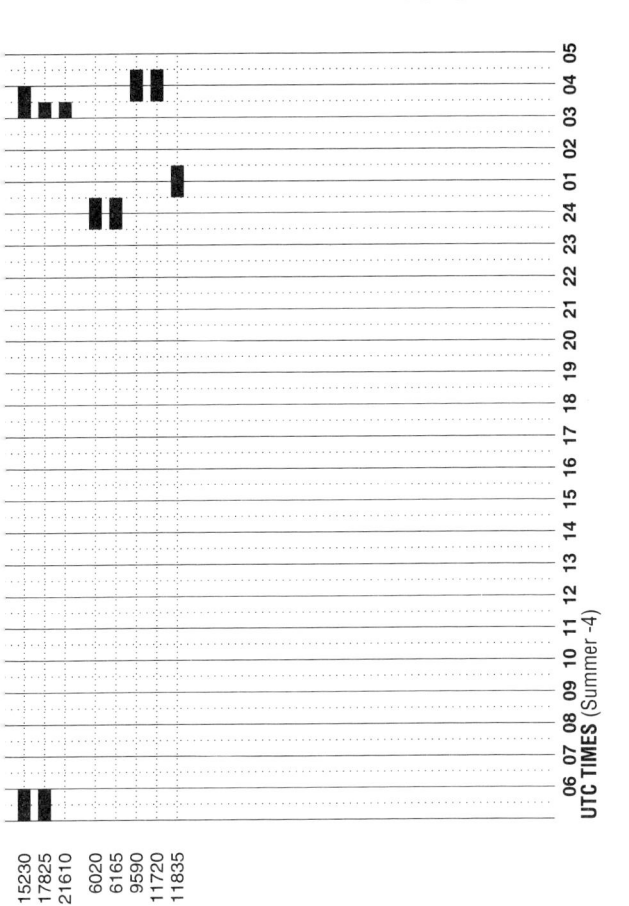

TORONTO

Radio Netherlands

Frequency	Schedule
15230	
17825	
21610	
6020	
6165	
9590	
11720	
11835	

UTC TIMES (Summer −4)

190 VIENNA

VIENNA, AUSTRIA
Hours difference from UTC (GMT): +1 (Summer +2)
Main business languages: German, English
Electricity: 50Hz, 220V
Currency: Schilling
Telephone country prefix and area code: 43 (1)

LOCAL AM & FM STATIONS	FREQUENCY
Blue Danube Radio	FM 92.9
Programs in English, French & German.	FM 103.8

INTERNATIONAL BROADCASTERS

BBC World Service
- 5965
- 6195
- 7120
- 7325
- 9410
- 9760
- 12095
- 15070
- 17640

Christian Science Monitor World Service
- 7510
- 9495
- 9840

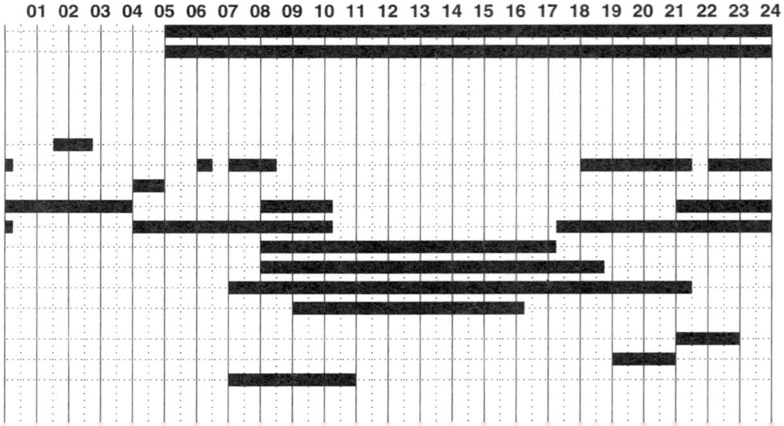

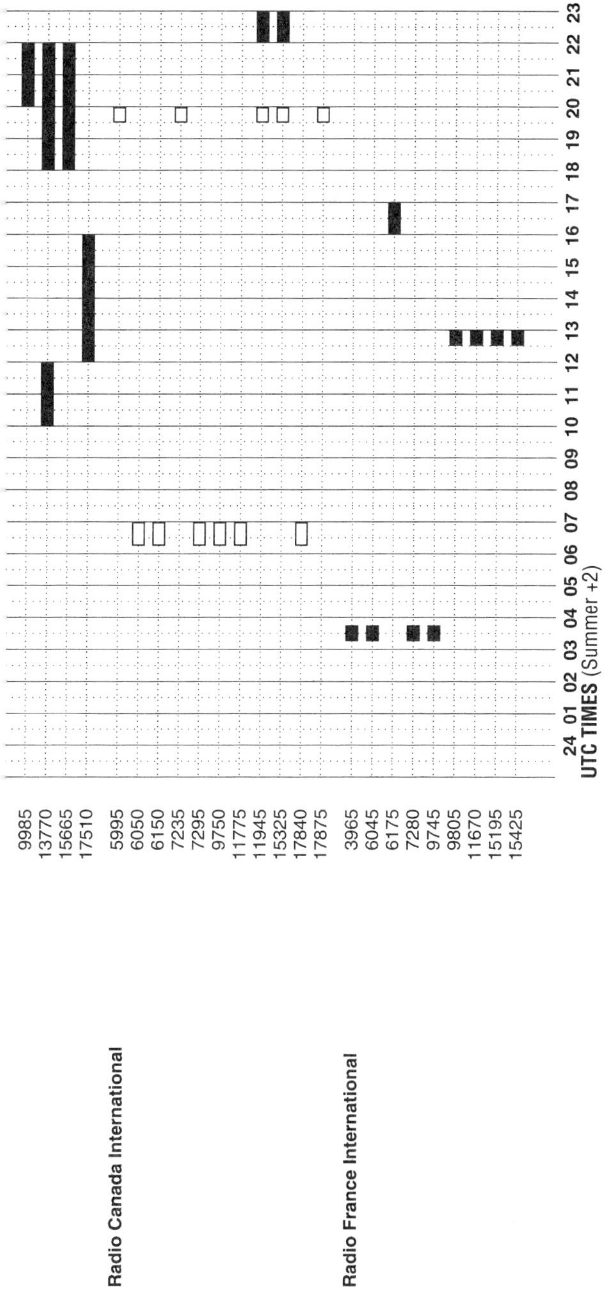

VIENNA

Radio Canada International

Frequency	Schedule
9985	
13770	
15665	
17510	

Radio France International

Frequency	Schedule
3965	
6045	
6175	
7280	
9745	
9805	
11670	
15195	
15425	

UTC TIMES (Summer +2)

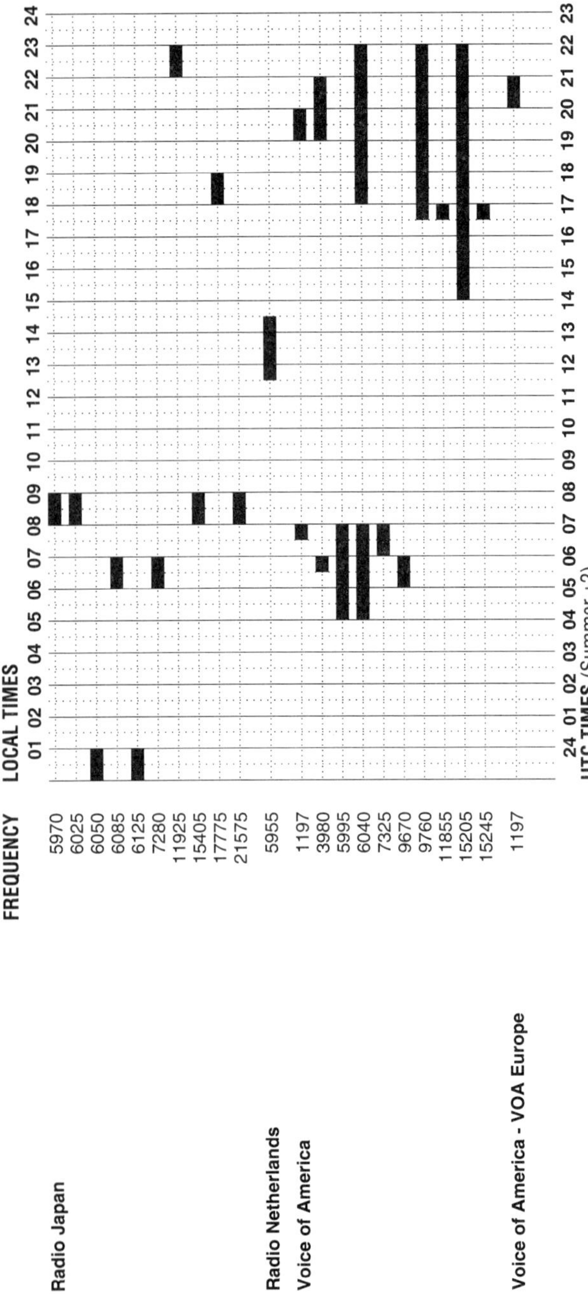

192 VIENNA

Broadcaster	Frequency (kHz)
Radio Japan	5970, 6025, 6050, 6085, 6125, 7280, 11925, 15405, 17775, 21575
Radio Netherlands	5955
Voice of America	1197, 3980, 5995, 6040, 7325, 9670, 9760, 11855, 15205, 15245
Voice of America - VOA Europe	1197

ZURICH 193

ZURICH, SWITZERLAND

Hours difference from UTC (GMT): +1 (Summer +2)
Main business languages: German, English
Electricity: 50Hz, 220V
Currency: Swiss Franc
Telephone country prefix and area code: 41 (1)

LOCAL AM & FM STATIONS

	FREQUENCY
VOA Europe	FM 91.0
Operated by the Voice of America. News on the hour. Available on hotel cable system only.	FM 104.6

INTERNATIONAL BROADCASTERS

BBC World Service

5965
6195
7120
7325
9410
9760
12095
15070
17640

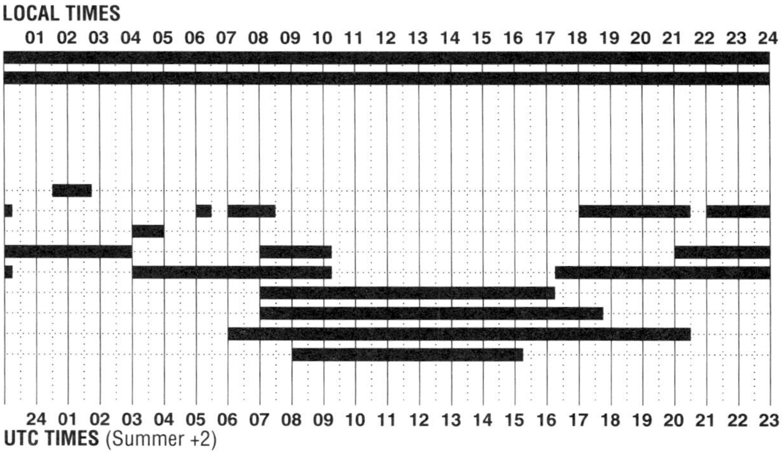

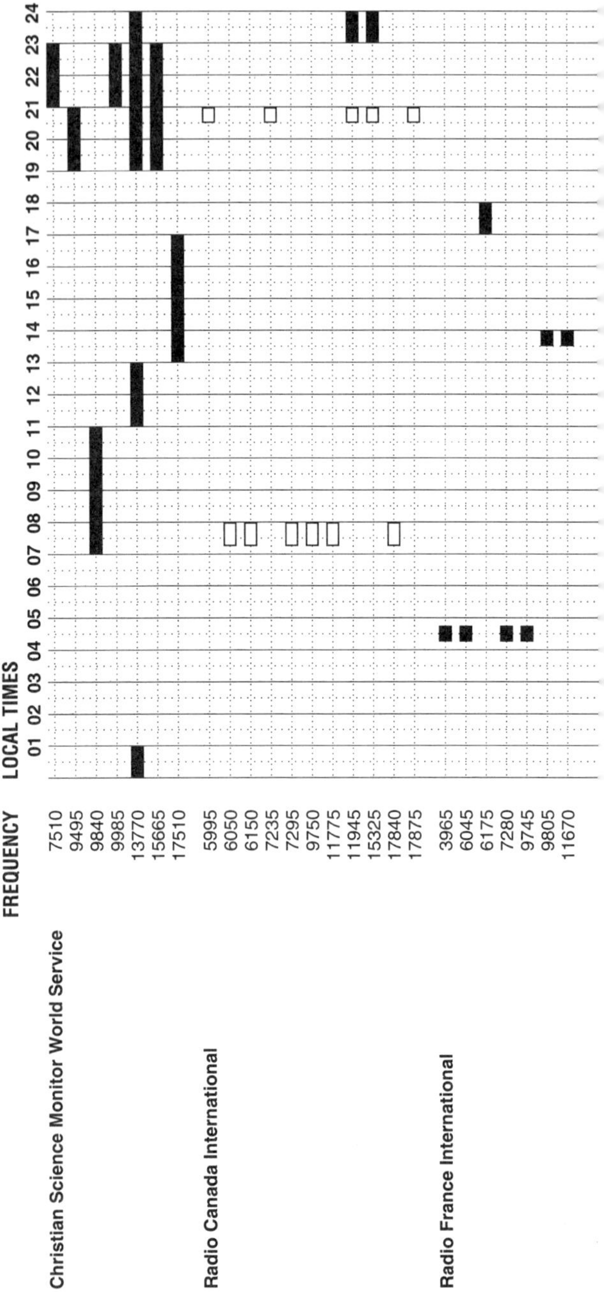

194 ZURICH

Christian Science Monitor World Service

Frequency	Schedule
7510	19-21
9495	
9840	07-10
9985	19, 22-24
13770	12, 19-24
15665	19-23
17510	00-01, 14-16

Radio Canada International

Frequency	Schedule
5995	07
6050	07
6150	21
7235	07
7295	07
9750	07
11775	21
11945	
15325	21
17840	21
17875	07

Radio France International

Frequency	Schedule
3965	04
6045	04
6175	17-18
7280	04
9745	04
9805	13-14
11670	13-14

ZURICH 195

Radio Japan
- 15195
- 15425

- 5970
- 6025
- 6050
- 6085
- 6125
- 7280
- 11925
- 15405
- 17775
- 21575

Radio Netherlands
- 5955

Voice of America
- 5995
- 6040
- 7325
- 9670
- 9760
- 11855
- 15205
- 15245

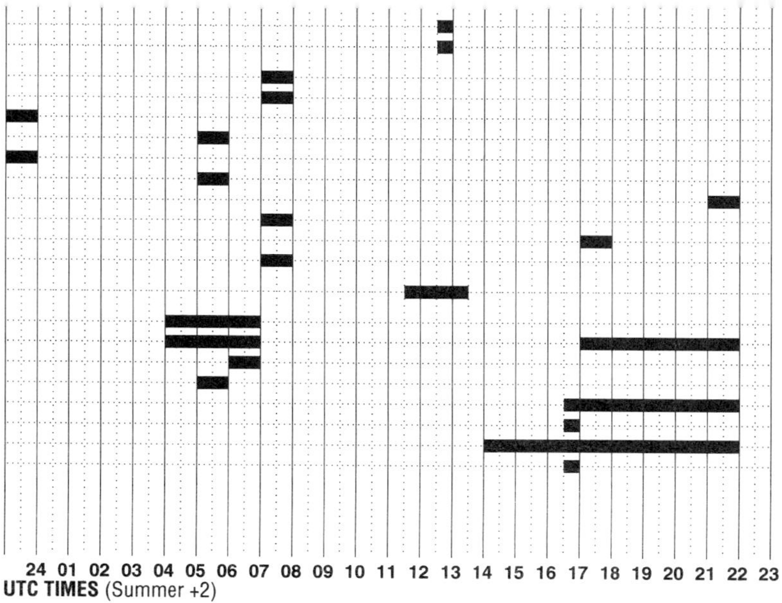

UTC TIMES (Summer +2)